Ecological Studies

Analysis and Synthesis

Volume 60

Ecological Studies

Volume 48
Ecological Effects of Fire in South African Ecosystems
Edited by P. de V. Booysen and N.M. Tainton
1984. XVI, 426p. 54 figures. cloth
ISBN 0-387-13501-4

Volume 49
Forest Ecosystems in Industrial Regions
Studies on the Cycling of Energy, Nutrients and Pollutants in the Niepolomice Forest, Southern Poland
Edited by W. Grodziński, J. Weiner, and P.F. Maycock
1984. XVIII, 277p. 116 figures. cloth
ISBN 0-387-13498-0

Volume 50
The Gulf of Agaba (Elat)
Ecological Micropaleontology
By Z. Reiss and L. Hottinger
1984. VIII, 360p., 207 figures. cloth
ISBN 0-387-13486-7

Volume 51
Soil Salinity under Irrigation
Processes and Management
Edited by I. Shainberg and J. Shalhevet
1984. X, 358p., 133 figures. cloth
ISBN 0-387-13565-0

Volume 52
Air Pollution by Photochemical Oxidants
Formation, Transport, Control and Effects on Plants
Edited by Robert Guderian
1985. XI, 346p., 54 figures. cloth
ISBN 0-387-13966-4

Volume 53
The Gavish Sabkha
A Model of a Hypersaline Ecosystem
Edited by G.M. Friedman and W. E. Krumbein
1985. X, 484p., 246 figures. cloth
ISBN 0-387-15245-8

Volume 54
Tidal Flat Ecology
An Experimental Approach to Species Interactions
By Karsten Reise
1985. X, 198p., 69 figures. cloth
ISBN 0-387-15447-7

Volume 55
A Eutrophic Lake
Lake Mendota, Wisconsin
By Thomas D. Brock
1985. XII, 308p., 82 figures. cloth
ISBN 0-387-96184-4

Volume 56
Resources and Society
A Systems Ecology Study of the Island of Gotland, Sweden
By James J. Zucchetto and Ann-Mari Jansson
1985. X, 248p., 70 figures. cloth
ISBN 0-387-96151-8

Volume 57
Forest Ecosystems in the Alaskan Taiga
A Synthesis of Structure and Function
Edited by K. Van Cleve, F.S. Chapin III, L.A. Viereck, C.T. Dyrness and P.W. Flanagan
1986. X, 240p., 81 figures. cloth
ISBN 0-387-96251-4

Volume 58
Ecology of Biological Invasions of North America and Hawaii
Edited by H.A. Mooney and J.A. Drake
1986. X, 320p., 25 figures. cloth

Volume 59
Acid Deposition and the Acidification of Soils and Waters
An Analysis
By J.O. Reuss and D.W. Johnson
1986. VIII, 120p., 37 figures. cloth
ISBN 0-387-96290-5

Volume 60
Amazonian Rain Forests
Ecosystem Disturbance and Recovery
Edited by Carl F. Jordan
ISBN 0-387-96397-9

Amazonian Rain Forests

Ecosystem Disturbance and Recovery

Case Studies of Ecosystem Dynamics Under a Spectrum of Land Use-Intensities

Edited by
Carl F. Jordan

Contributors
Robert J. Buschbacher, Carl F. Jordan, Charles E. Russell,
Juan G. Saldarriaga, Geoffrey A.J. Scott, E.A.S. Serrão,
Christopher Uhl

With 55 Figures

Springer-Verlag
New York Berlin Heidelberg
London Paris Tokyo

Carl F. Jordan
Institute of Ecology
University of Georgia
Athens, Georgia 30602
USA

Library of Congress Cataloging in Publication Data
Amazonian rain forests.
(Ecological studies; v. 60)
Includes index.
1. Deforestation—Amazon River Region—Case studies.
2. Rain forest ecology—Amazon River Region—Case studies. 3. Shifting cultivation—Amazon River Region—Case studies. 4. Forests and forestry—Amazon River Region—Case studies. 5. Agricultural ecology—Amazon River Region—Case studies. I. Jordan, Carl F. II. Series.
SD418.3.A53A53 1986 634.9′0981′1 86-21962

Typeset by TC Systems, Shippensburg, Pennsylvania.
Printed and bound by Quinn-Woodbine, Woodbine, New Jersey.
Printed in the United States of America.

9 8 7 6 5 4 3 2 1

ISBN 0-387-96397-9 Springer-Verlag New York Berlin Heidelberg
ISBN 3-540-96397-9 Springer-Verlag Berlin Heidelberg New York

Acknowledgments

The studies reported in Chapters 2, 3, and 5 were supported by Ecosystem Studies Program of the U.S. National Science Foundation (NSF); CONICIT de Venezuela (Venezuelan Science Foundation); the Organization of American States; Institute of Ecology, University of Georgia; Centro de Ecologia, Instituto Venezolano de Investigaciones Cientificas (IVIC); and the UNESCO Man and the Biosphere Program, which also supported work in Chapter 8. NSF also funded the study in Chapter 4. The Chapter 7 study was assisted by Jari Florestal e Agropecuaria and the New York Botanical Garden.

Contents

Contributors

Robert J. Buschbacher	Institute of Ecology University of Georgia Athens, Georgia, USA
Carl F. Jordan	Institute of Ecology University of Georgia Athens, Georgia, USA
Charles E. Russell	Caesar Kleberg Wildlife Research Institute Texas A & I University Kingsville, Texas, USA
Juan G. Saldarriaga	Oak Ridge National Laboratory Environmental Sciences Division Oak Ridge, Tennessee, USA
Geoffrey A.J. Scott	Geography Department University of Winnipeg Winnipeg, Manitoba, Canada

E.A.S. SERRÃO

Centro de Pesquisa Agropecuária do Trópico Úmido
EMBRAPA
Belém, Pará, Brazil

CHRISTOPHER UHL

Department of Biology
The Pennsylvania State University
University Park, Pennsylvania, USA

1. Introduction

DEVELOPMENT AND DISTURBANCE IN AMAZON FORESTS

Contrasting Impressions

The rain forests of the Amazon Basin cover approximately 5.8×10^6 km^2 (Salati and Vose 1984). Flying over even just part of this basin, one gazes hour after hour upon this seemingly infinite blanket of green. The impression of immensity is similar when viewed from the Amazon River itself, or from its tributaries. From a hammock on the shaded deck of a riverboat, the immensity of the forest presents an incredible monotony as one view of the shoreline blends unnoticeably into another. From both perspectives, the overwhelming reaction to the sea of trees that stretches from horizon to horizon is a sense of the vastness of the rain forest.

In September 1985, I got a different impression of the rain forest. Several students and I journeyed in a self-propelled car along the single-track railroad that stretches almost 1000 km from the Carajas iron ore mine in the rain forest of Pará State, Brazil, all the way to São Luis on the coast (Fig. 1.1). The railroad had been built through virgin rain forest for almost the entire length, except for the last stretch near the coast where there was a transition to dry or scrub forest. The railroad was completed in February 1985. By the time I took the trip, half to two-thirds of the forest along the railroad had been cleared between several hundred meters to several kilometers on either side of the track. A striking view along almost the entire length of the track was

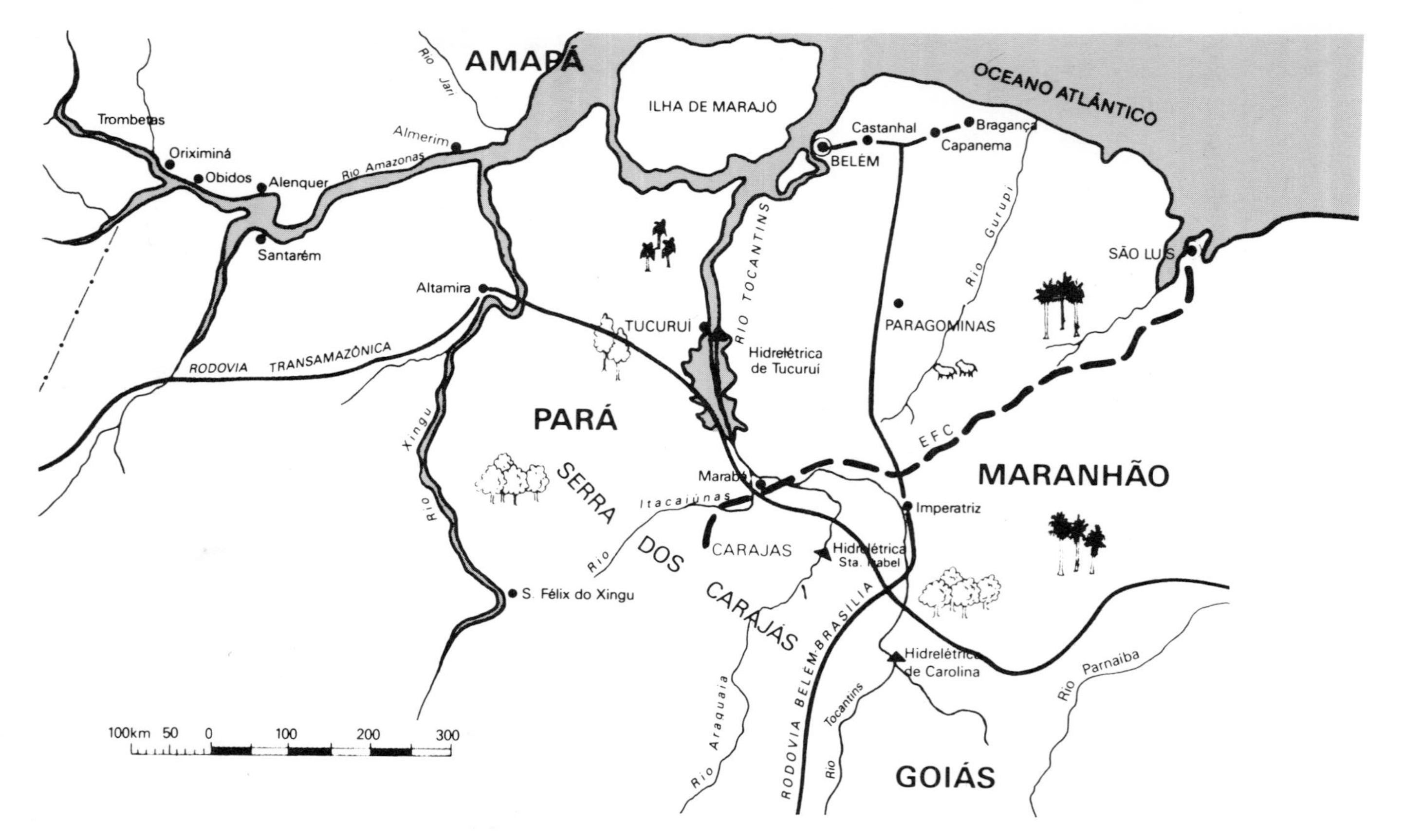

Figure 1.1. Map of the eastern portion of the Amazon Basin. The mouth of the Amazon River is in the top center. The railroad from the iron mine at Carajás to São Luis is indicated by the dashed line.

that of smoke rising from fires that were burning the dried remains of cut forest.

Disappearance of the Rain Forest

Because the Amazon rain forest is so huge, the cutting along the railroad and along other roads penetrating the region might not seem very important. It seems that plenty of the forest remains. However, this impression of a wilderness without limits is the same illusion that has deceived pioneers and developers in other forested regions of the world. For example, in the mid-1800s the huge forests that covered Wisconsin and Michigan, which were cut to build midwestern United States, seemed inexhaustible (Twining 1983). In the words of one old lumberjack, "Us loggers thought the big woods would last forever" (Fritzell 1983). Yet, by the early 1900s, the huge pine forests were gone, with the exception of a few inaccessible patches (Reinhardt 1983). And the clearing of the midwestern forests was accomplished without power saws and power machinery, now frequently used in Amazonia. The lesson from history is that the apparent infinity of the forest is an illusion. The clearing of the Amazon could take less time than the clearing of the midwestern United States.

Cutting is not the only cause of rain forest destruction. Fires that escape from fields into surrounding forest also cause damage. For example, the February 12, 1986 issue of *Veja,* a Brazilian weekly news magazine, and the February 6, 1986 issue of *Jornal Do Brasil* reported satellite detection of a forest fire in Amazonia that covered an area 400 by 600 kilometers.

Although there is some disagreement about the exact rates at which the Amazon rain forests are being destroyed (Myers 1980, Lugo and Brown 1982), most scientists concede that by the end of this century much of the primary forest in the eastern Amazon will probably be gone, and much of the western Amazon will be affected by man.

Reasons for Deforestation

Why are the Amazonian rain forests being cut? There are a number of reasons:

1. Population pressure is great in many coastal and mountain regions of countries having territory in the Amazon Basin. Transmigration of landless peasants and jobless urban poor into the sparsely settled Amazon is one temporary political solution for overpopulation.
2. Most of the countries with land in the Amazon need goods to export to reduce their foreign debt, and the rain forest, or food and fiber derived from converted rain forest ecosystems, can supply some of that capital.
3. There is a feeling in governments of some Amazonian countries of a need to defend the sparsely populated Amazon border regions against encroachment by citizens of the adjoining country (Wagley 1984).

Is the Rainforest Used Wisely and Well?

The rapid deforestation in the Amazon Basin raises questions about whether the land is being used wisely and well or is being wasted. Although in many areas of the rain forest native Indians still practice the shifting cultivation that has sustained them for thousands of years, these practices are being replaced rapidly with other types of agriculture. In some regions of the Amazon, huge areas of the forest are cut and pulpwood or pasture is cultivated. In other regions, landless peasants are migrating into the forest and cutting it so they can plant food crops. Many of these efforts at rain forest development have short-lived results. For example, along the roads that penetrate the rain forest the most common view is of small secondary vegetation, evidence that the original forest has been cut and the land perhaps cultivated, or grazed, and then abandoned (Smith 1982).

During my railroad trip through Pará in Amazonia, I noticed that near towns, logs from clear-cut forests sometimes were hauled to local sawmills. However, much of the clearing was for pasture, and cut trees were simply burned in place. Pasture establishment purportedly is for raising beef, but a more important reason may be that it is a mechanism for establishing proprietary rights over large areas of land, as was the case in eastern Pará near Paragominas (Hecht 1984). Where the land is cleared only to establish and hold a claim and not to produce crops, the cutting and burning certainly appear to squander a forest resource.

Arguments for Conservation

The apparent haste and wastefulness of much current development in the Amazon has prompted calls for slowing the losses of species, lessening the land degradation, and preventing other negative effects of rapid deforestation (De Melo Carvalho 1984).

Amazon rain forests deserve special environmental attention because they are one of the most biologically diverse ecosystems in the world (Prance 1982). Arguments have been made for preserving species on the basis of their potential economic value. For example, Myers (1984) has pointed out that there are economic and health benefits from foods and pharmaceuticals derived from plants indigenous to the rain forest, and that there are potentially many more products from forest species, but that these species will become extinct if clear-cutting continues unabated. In addition, continued deforestation may decrease the rainfall in the Amazon Basin (Salati and Vose 1984), thereby eliminating even more species adapted to the rain forest environment.

These are utilitarian arguments for species preservation, but it can be argued from an ethical point of view that species have a right to exist regardless of whether or not they happen to be useful to man (Rolston 1985). Moral arguments also are used to justify preservation of the forest so that native Indian tribes can continue their existence and culture unaffected by changes caused by development (Davis 1977).

Arguments for conserving Amazonian forests also have been based on the low potential of Amazonian soils for producing agricultural crops (Goodland and Irwin 1975). It is well known that many soils in Amazonia are very low in nutrients and that the Amazon Basin contains a greater proportion of low-fertility soils, such as Oxisols and Ultisols, than any other region on earth (Sanchez 1981). Such lands should be considered marginal for agriculture (Sioli 1973), just as desert and mountain soils usually are considered marginal for crops in other regions.

Objective

Despite the arguments of conservationists, it is clear that the Amazon rain forest will not be preserved in a pristine condition. The pressures for development are too great. However, the environmental destruction that has accompanied many development projects may not be necessary. There may be other ways of using the Amazon forest that are better. Development may be accomplished in a more ecologically sound way.

Even though much of the Amazon rain forest will soon be developed, there still is time to evaluate development practices and change them when it is clear that there are better alternatives. The objective of this book is to examine approaches to rain forest development in the Amazon through an examination of case studies of forest and land utilization. A comparison of the case studies in the last chapter evaluates the approaches to development according to their social, economic, and political aspects, as well as from the environmental viewpoint.

Case Studies

The case studies are of man-caused perturbations that have been, or are becoming, common in the Amazon rain forest region. They are arranged and presented in an order. They begin with disturbances that are small in scale, short in duration, and mild in intensity. They end with disturbances large in scale, relatively long in duration, and severe in intensity.

The cases are: shifting cultivation where land is unlimited (slash and burn agriculture near San Carlos de Rio Negro, Venezuela); recovery following shifting cultivation in the upper Rio Negro (a chronosequence of abandoned shifting cultivation sites); shifting cultivation where land is limited (Campa Indian agriculture in the Gran Pajonal of Peru); pasture as a claim to land (government-sponsored development in the Amazon Territory of Venezuela); permanent plots and continuous cropping or harvesting (colonization along the Trans-Amazon Highway; intensively fertilized crops at Yurimaguas, Peru; Agroecology at Tome-Assú, Brazil; sustained-yield forestry in Suriname); plantation forestry (the Jari project, Pará, Brazil); large-scale pasture development (abandoned pastures near Paragominas, Brazil). Site locations are shown in Figure 1.2.

Case studies include sites that may be classified as lowland tropical rain forest, tropical moist forest, or evergreen seasonal forests. The term "Amazon rain forests," used generically in the title of this book, may not be appropriate according to some forest classification schemes. Sites are on soils classified as Ferralsols by the FAO system (Latosols, Oxisols, Ultisols, and similar types in other systems, and often locally called *Terra firme*). Sites on podzol, alluvial soils, and mountain soils are excluded. One study site is in Suriname, outside the drainage basin of the Amazon but still within the Amazon type of forest.

Selection of Case Studies

Two of the case studies were doctoral dissertations for which I was the major professor (pastures in the upper Rio Negro, Chapter 5; Jari, Chapter 7), and one was a dissertation for which I was an advisory committee member (chronosequence following shifting cultivation, Chapter 3). One case study (shifting cultivation, San Carlos, Chapter 2) was a research project for which I was a principal investigator. The large-scale pasture development study (Chapter 8) was carried out by a former student and postdoctoral associate. This book grew out of the idea that it would be worthwhile to condense these dissertations and project reports and to compare them in a single, easily accessible volume.

The studies were similar in many respects. They all dealt primarily with nutrient cycling and productivity in the Amazon rain forest and changes in these parameters as a result of man-caused disturbances. All the studies employed similar methods. The major differences among the studies were the intensity and scale of the disturbances. Comparison of the case studies is analogous to an experiment in which replicated systems are subjected to increasing levels of stress to determine response as a function of stress.

After work on the book began, another dissertation was brought to my attention that dealt with an Amazonian ecosystem and that was very similar in approach and methodology to the cases already considered, so this dissertation was included (shifting cultivation in the Gran Pajonal, Chapter 4).

As work on the book progressed, it became apparent that two ends of the disturbance spectrum were covered by case studies, but that the intermediate range of disturbance was not represented. Because the published literature contains studies of permanent plots for agriculture and forestry in the Amazon, I summarized several of these cases (Chapter 6) to complete the spectrum of disturbance intensities. One case was the Trans-Amazon colonization project; a second was of intensive fertilization of continuously cultivated annual crops; a third dealt with management of native forests; and a fourth was the Japanese labor-intensive cropping system at Tome-Assú, Brazil. The Tome-Assú case was included not because data were available but because, on the basis of a

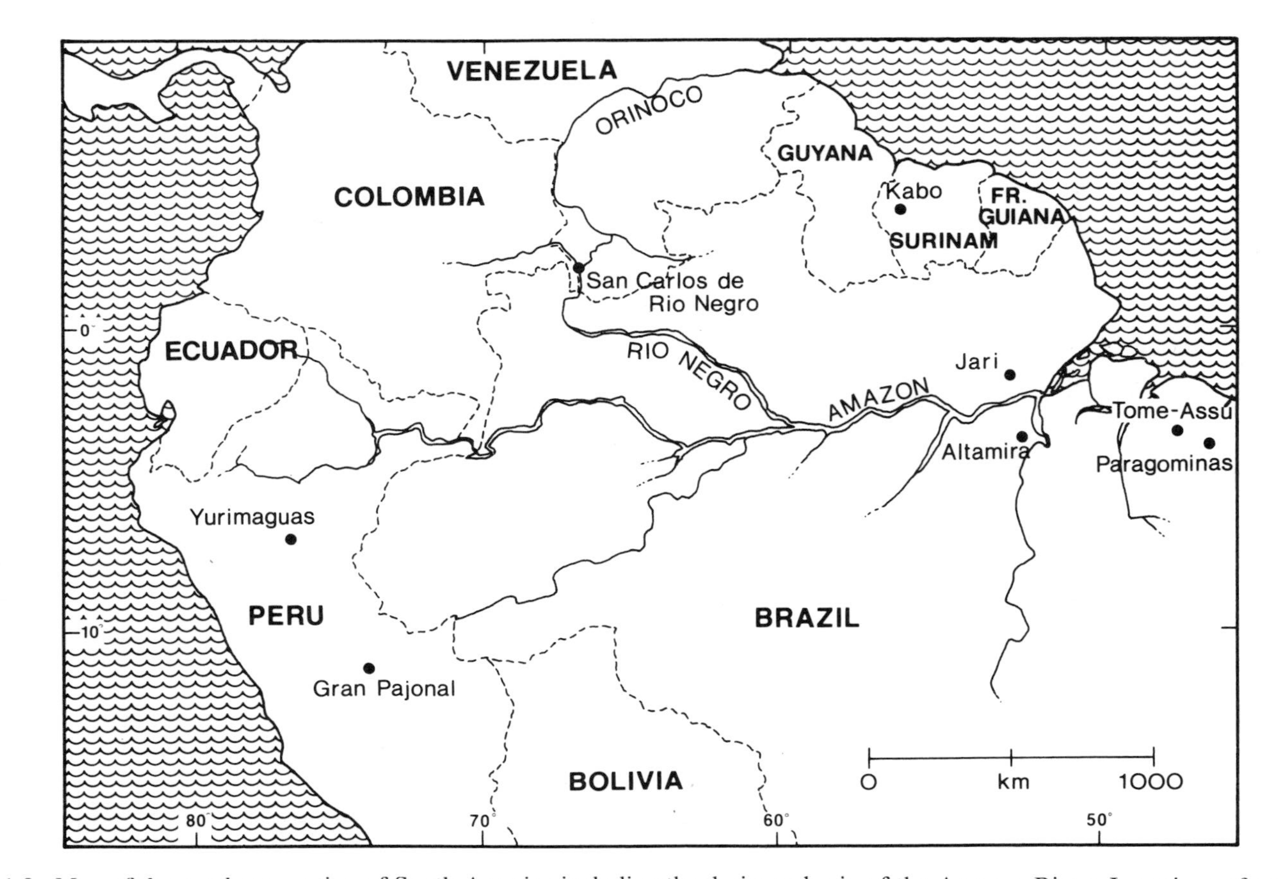

Figure 1.2. Map of the northern portion of South America including the drainage basin of the Amazon River. Locations of case study sites are shown with dots.

visit, it subjectively appeared to be a continuous cropping system that was both productive and sustainable without large fertilizer inputs.

The final chapter compares the case studies and assesses, to the extent possible, the economic profit, the social stability, the political success, and the environmental changes resulting from each situation.

Slowing Deforestation

The primary motive for putting together this book arises from my observations of the environmental destruction that is accompanying development in Amazonia. Is this destruction inevitable? Can the rate of destruction be slowed?

The rate at which the rain forest is cut, and the amount that can be conserved, depend in large part on how wisely and well the exploited areas are used. Some methods of utilization are much less wasteful of the forest resource and are more productive than other methods. Better ways of using some of the forest leads to conservation of the remainder. It is hoped that the comparisons of case studies in this book illustrate some of the strengths and weaknesses of each approach.

2. Shifting Cultivation*

Case Study No. 1: Slash and Burn Agriculture Near San Carlos de Rio Negro, Venezuela

Carl F. Jordan

The term "shifting cultivation" refers to farming or agricultural systems in which a short cultivation phase on land that has been cleared and burned alternates with a long fallow period. The cultivation technique is often referred to as "slash and burn agriculture." Some of the local names that have been given to this type of agriculture in South America are *roça* in Brazil, *conuco* in Venezuela, and, *chacra* in Peru (Savage et al. 1982).

To initiate slash and burn agriculture, the forest, secondary bush, woodland or grassland vegetation is cleared with simple hand tools, and herbaceous plants, vines, and saplings are cut or "slashed" (Okigbo 1984). Useful trees and shrubs are sometimes left standing with light pruning. Others may be pruned down to stumps of varying height for fast regeneration or for support of vine crops. The dry plant debris is then burned in regions where the dry season is sufficiently long; but in regions where frequent rains make burning difficult, such as the Atlantic lowland rain forest area of Costa Rica, planting may take place without it (personal observation, 1981).

Burning reduces stump sprouting, which competes with crops. Burning usually does not consume large trunks but it eliminates smaller branches and

---

* Except where otherwise cited, descriptions, data, and findings in this chapter are from: Jordan C. F. 1975–1984. Nutrient dynamics of a tropical rain forest ecosystem, and changes in the nutrient cycle due to cutting and burning. Annual reports submitted to the U.S. National Science Foundation. Institute of Ecology, University of Georgia, Athens, Georgia.

makes the field more accessible for planting. It may kill seeds of weed species in the surface soil and also may affect soil bacteria (Nye and Greenland 1960). A quick pulse of nutrient-rich ash to crops is often regarded as an important benefit of burning (Nye and Greenland 1960). However, the case study reported here indicates that the slow release of nutrients from unburned decomposing organic matter on the soil surface may be more important in sustaining production over the 2- or 3-year cropping period.

In shifting cultivation, crops often are grown in mixtures of annual and perennial food crops, and tree crops. Organic matter, such as litter and carcasses from the surrounding forest and streams, may be applied to the soil, but commercial fertilizers are rarely used. Almost all the products of slash and burn agriculture are consumed by the cultivator and his family or tribe, with only a small proportion being traded for metal tools and other implements. The crops planted, the cultivation techniques used, and the length of fallow vary greatly among regions, and even among sites within regions (de Schlippe 1956, Watters 1971, UNESCO 1983).

The cultivation phase alternates with a much longer fallow period, in which naturally occurring secondary successional vegetation invades the plot. In some cases, trees valuable for fruit or wood are planted as fallow (Denevan et al. 1984). Fallow should be long enough to allow sufficient rebuilding of nutrient stocks for another rotation of crops. Where population densities are low, adequate fallows are usually possible. Such agriculture is ecologically sustainable, because the fallow period restores nutrients lost during cultivation and harvest.

Evolution of Shifting Cultivation

Several thousand years ago, the inhabitants of many tropical forests were basically hunters and gatherers. Gradually there evolved a practice among some tribes of protecting and then even planting of species that produced products favored for food or medicinal uses. Many of the favored crop species grew better in cleared open areas, and consequently there evolved the practice of opening clearings in the forest where these species were planted (Meggers 1971, 1985).

The first tools for cutting and clearing were stone axes, and these primitive tools may still be used by some tribes isolated in remote rain forest regions. However, within the past several hundred years, most of the indigenous tribes have had at least some contact with colonists, missionaries, explorers, and other representatives of technologically advanced societies. This contact has resulted in the first step of "development" for the previously isolated tribes and the forest that they inhabit (Miracle 1973). Often, this early contact between "developers" and "natives" included trade of steel axes and machetes (cutlasses) for natural plant and animal products, such as food and medicines, and for such products as woven baskets in return for labor and services of the native people or for becoming Christianized (Wilbert 1972).

Despite contact with developed culture, many groups of native peoples continued to use the forest in the same manner that they had for generations past. The culture of the people and their daily habits were intertwined with the processes of forest clearing, planting, and fallowing (Meggers 1971). Where agricultural ritual has become part of the cultural tradition, agriculture may be called "people centered" or "culturally centered" (Korten and Klauss 1984).

As long as tribes remain relatively isolated, substitution of steel for stone cutting tools usually does not change the basic method of cultivation, or the crops used, when tribes come in contact with a market economy. It just makes the work easier. Most of the daily life of the cultivators changes very little. However, as exchange with a market economy grows, native peoples often seem eager to adopt developed technology (Gross et al. 1979).

Shifting Cultivation in the North-Central Amazon Basin

The Orinoco River originates in the mountains along the eastern border of the Amazon Territory of Venezuela, flows westward out of the mountains, and then turns northward toward the plains, or Llanos, of Venezuela. Before it begins its major turn northward, it splits into two parts, with part of the river flowing southward through the Casiquiare canal, which joins the Rio Negro near San Carlos, Venezuela (Fig. 1.2). Because the Rio Negro then continues on and joins the Amazon, the central and southern portion of Venezuela's Amazon Territory are within the drainage basin of the Amazon River.

Forms of shifting agriculture and cultivation of food and medicinal crops have been carried out in Venezuela near the upper Orinoco for at least 3000 years, by a group of Carib-speaking tribes now known as the "Ye'cuana" or variously as "Makiritare" (Wilbert 1972, Key 1979). Early practice was mainly forest gardening, in which naturally occurring individuals of food bearing trees were protected and there was some cultivation of manioc or yuca (*Manihot esculenta* Crantz). Stone axes were used to clear forest for planting of manioc. Today, yuca still is the staple crop, but tribes also cultivate plantains, bananas, maize, sugarcane, pineapples, papaya, various root crops, and several medicinal plants (Wilbert, 1972). Machete and steel axes now are usually available for forest clearing.

The Ye'cuana occupy parts of the north and central portion of the Amazon Territory of Venezuela. Figure 3.1 (Chapter 3) shows the relationship of the Amazon Territory to the rest of Venezuela. The southeastern portion of the territory, and across the Sierra Parima into Brazil are the grounds of the Yanoama (Yanomamo) tribes, which are primarily a hunting and gathering society. Most of the Yanoama subtribes adopted horticulture only within the past hundred years (Wilbert, 1972). This may have resulted from recent increased peaceful interactions with the Ye'cuana.

In the southwestern portion of the Amazon Territory, in the region where the Casiquiare and Guainia rivers unite to form the Rio Negro, Arawak-speaking tribes, such as the Wakuenai (Curripaco), are the predominant groups (Lathrap 1970, Key 1979). Their horticultural systems are indigenous to the "blackwater ecosystems" of the Guainia drainage area, so-called because of the dark but clear amber color of the streams which drain the area. Fertility of soils in this region is very low, and a large effort is devoted to fishing to supplement the diet (Moran and Hill 1982).

In the Amazon Territory of Venezuela near the village of San Carlos de Rio Negro are remnants of the Arawak-speaking Bare and Baniwa tribes. Many of the descendants live in or near the village and cultivate yuca nearby. The techniques of cultivation may be influenced strongly by Arawak traditions for cultivating soils of very low fertility. However, they also may be influenced by the Yanoama, because soils of sites selected for cultivation are usually Oxisols or more fertile Ultisols (See Case Study No. 2, Chapter 3). These soils are similar to the more common soil types in the Yanoama territory at higher elevations to the east (R. Herrera, personal communication, 1977).

In the area around San Carlos, root of "sour" yuca is the principal crop. The sour variety is preferred for subsistence farming because of its resistance to herbivory and decomposition. The root can be left in the ground until it is convenient to harvest. Propagation is by cuttings of mature plants, which are inserted in small holes made with a "dibble stick." Extraction of hydrocyanic acid, which is a naturally occurring chemical defense of the sour yuca, is usually carried out in the traditional manner, through grating, soaking, and squeezing of the tuberous roots (Uhl and Murphy 1981). Processing is accomplished with the use of implements crafted from the forest, although tin cut into fine points now sometimes replaces quartz as the shredders on the grating board.

As is the custom throughout the Territory, land outside the village is not "owned" in the sense of a title that can be bought and sold. We learned this through talking with the prefecto, or mayor, of San Carlos. The land belongs to whomever is cultivating it and using it. An individual who wants to establish his own cultivated site, or *conuco* as they are locally called, explores the forest until he finds a site deemed suitable. Cultivators are careful not to locate their *conuco* too close (several hundred meters) to an already established *conuco,* to allow for possible future expansion.

The San Carlos Experiment

Between 1974 and 1983, nutrient cycling and productivity were studied in a *conuco* near the village of San Carlos de Rio Negro. San Carlos is in the southern part of Venezuela's Amazon Territory, near the common boundary of Venezuela, Brazil, and Colombia (Fig. 1.2). The study was part of a scientific project coordinated through Instituto Venezolano de Investigaciones Cientificas (IVIC), Caracas, Venezuela, to develop an understanding of the structure and functioning of an Amazonian rain forest (Herrera et al. 1981).

Site Description

Much of the terrain surrounding San Carlos is flat, and the soils are coarse podzolized sands supporting a forest of relatively small structure, called Amazon Caatinga. Interspersed in the flat terrain are "islands," usually of a few hectares and extending up to 40 m above the surrounding lowlands. In this region, the hill soils are usually Oxisols and Ultisols (Dubroeucq and Sanchez 1981), and the forests are larger and have greater species diversity than the Caatinga forests. Usually only Oxisols and Ultisols are used for slash and burn agriculture near San Carlos.

A characteristic of the undisturbed forest that is important in nutrient conservation is a mat of litter and coarse humus mixed with roots on top of the soil surface. On the Oxisol sites, the mat ranges in thickness from just a few to 40 cm or more in thickness. A large proportion of the nutrients entering the mat by way of litterfall or throughfall are taken up by the roots before they can leach down into the mineral soil. The tight coupling of decomposing litter with root uptake appears to decrease the potential for losses by leaching, volatilization, or fixation by the mineral clays.

An experimental plot of 1 ha was located on one of the Oxisol hills, and a similar control plot was established about 50 m away on the same hill. Forest biomass and standing stock of nutrients were determined in both plots before the experimental plot was cut. The two plots differed very little in biomass (Fig. 2.1). Greatest stocks of calcium and potassium were in the above-ground biomass, but greater stocks of nitrogen and phosphorus were below ground (Fig. 2.1). Of the total soil stocks of phosphorus, only about 2% were readily soluble.

Methods

The primary forest on the experimental plot was cut in September 1976. Forest cutting and burning, planting, weeding, and harvesting were carried out by local workers under the supervision of an experienced local farmer, to insure that operations conformed with local practice. The forest understory was first cleared out with machetes. Then, a number of small to medium-sized tress that were linked by canopy lianas to a large tree were partially cut through. The large tree was then cut completely, and as it fell it pulled down the smaller trees. The felled trees dried for about 4 months and were burned in December 1976. Cultivation began in January 1977. The main crop planted in the *conuco* was sour yuca. Interspersed between the yuca plants were pineapple, plantain, and cashew, which is a tree crop.

Planting and harvesting in a *conuco* is an almost continuous process. Planting of the second crop of yuca begins before the first crop is harvested. Pineapple begins to bear fruit the second year, and plantain is generally fruiting by the second or third year, although it did not fruit during 3 years in this plot. The final harvest of yuca was completed in January 1980, and the plot was aban-

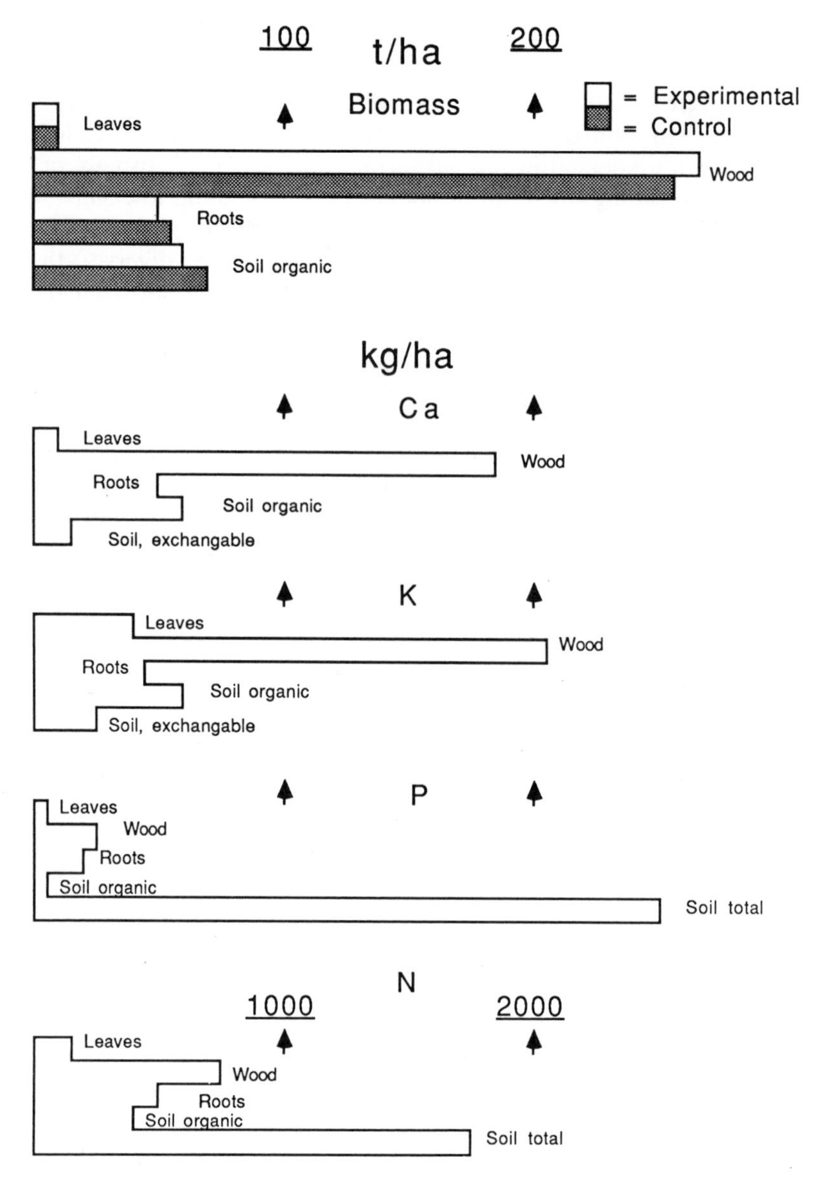

Figure 2.1. Dry weight biomass and nutrient stocks of forest and their distribution among structural components in the primary forest at the beginning of the experiment. "Soil organic" includes fallen tree trunks and above- and below-ground humus.

doned from cultivation. However, the cashew trees were just beginning to fruit, and the nuts were available for several more years.

This cropping system had only a few species compared to *conucos* along the upper Orinoco, probably because the soil near San Carlos is lower in nutrient content. Nutrient-demanding crops, such as corn and sugarcane, sometimes planted in the upper Orinoco, cannot grow around San Carlos. Even banana and plantain do poorly near San Carlos, as evidenced by the failure of plantain to fruit in the experimental plot.

Standing stock of biomass and nutrients in the experimental *conuco* were measured at intervals of 1 year or less, from the time of cutting, through July 1983, 3½ years after abandoment of cultivation. Nutrient input and loss from the plot and primary productivity were measured continually during the period of cultivation and less intensively after abandonment.

Results and Discussion

Nutrients

Leaching Losses

In the San Carlos experiment, there was a sharp increase in leaching losses of calcium, magnesium, potassium, and nitrate nitrogen following cutting and burning (Fig. 2.2). However, ammonium nitrogen did not differ between the experimental and the control sites during the experiment. This may have resulted from a rapid increase in nitrifying bacteria in the experimental plot following cutting and the conversion of ammonium nitrogen to nitrate before the ammonium could be leached. There was no detectable increase in phosphate phosphorus in the leachate water during the course of the experiment, probably because the soluble phosphate was fixed by iron and aluminum in the mineral soil before the phosphate could be leached out of the system.

The sharp increase in leaching losses after cutting and burning was expected. Studies of nutrient leaching in North America have shown that there almost always is a sharp increase following clear-cutting. Research at Coweeta Hydrologic Laboratory in North Carolina showed higher rates of nitrate loss and sediment loss in recently cut catchments than in control catchments (Monk 1975, Swank and Douglass 1975, 1977, Webster and Patten 1979). Studies of nutrient dynamics in conifer forests of the Pacific Northwest also showed an increase in loss rates following clear-cut harvesting and burning of remaining slash (Gessel and Cole 1965, Miller and Newton 1983, Feller and Kimmins 1984).

In a study on nutrient loss following clear-cutting carried out in northern hardwoods at the Hubbard Brook watershed site in New Hampshire, Bormann et al. (1968) concluded that

> clear cutting tends to deplete the nutrients of a forest ecosystem by reducing transpiration and so increasing the amount of water passing through the system;

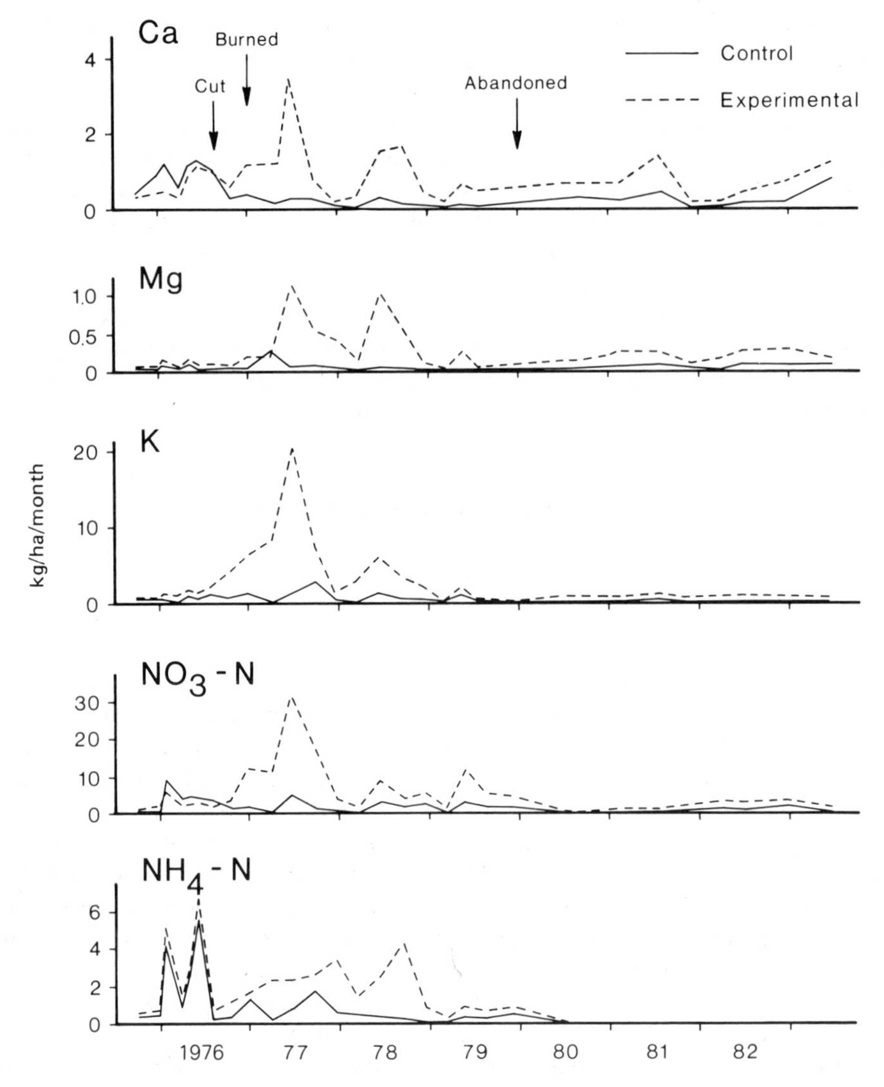

Figure 2.2. Monthly rates of nutrient leaching in the control and cultivated plots.

simultaneously reducing root surfaces able to remove nutrients from the leaching water; removal of nutrients in forest products; adding to the organic substrate available for immediate mineralization; and in some instances, producing a microclimate more favorable to rapid mineralization. [p. 884]

Nutrient Stocks

The stocks of calcium, potassium, magnesium, and nitrogen in the experimental plot throughout the study are shown in Figure 2.3. The compartments of the ecosystem, such as the soil, trunks, etc., as labeled in the legend are stacked.

The total at the top of the graph at any point along the x axis is the sum of stocks of that nutrient in all compartments of the ecosystem at that time. Below zero on the y axis is the cumulative loss of nutrients as a function of time. Cumulative loss results from excess leaching over atmospheric deposition, excess harvest over importation of planting stock, and excess denitrification over nitrogen fixation. Measurements of cumulative net losses resulting from

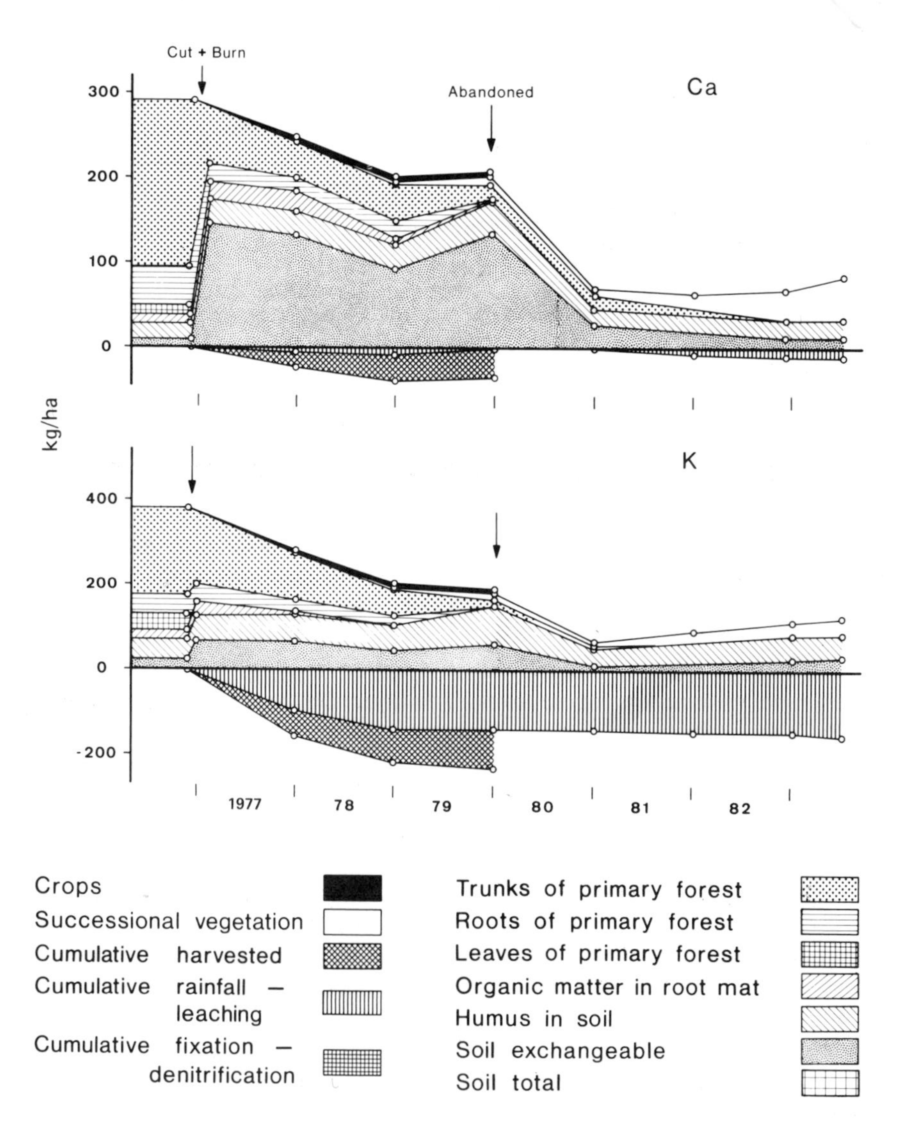

Figure 2.3. Stocks and cumulative losses of calcium, potassium, magnesium, and nitrogen as a function of time in the experimental plot. See key and text for further explanation.

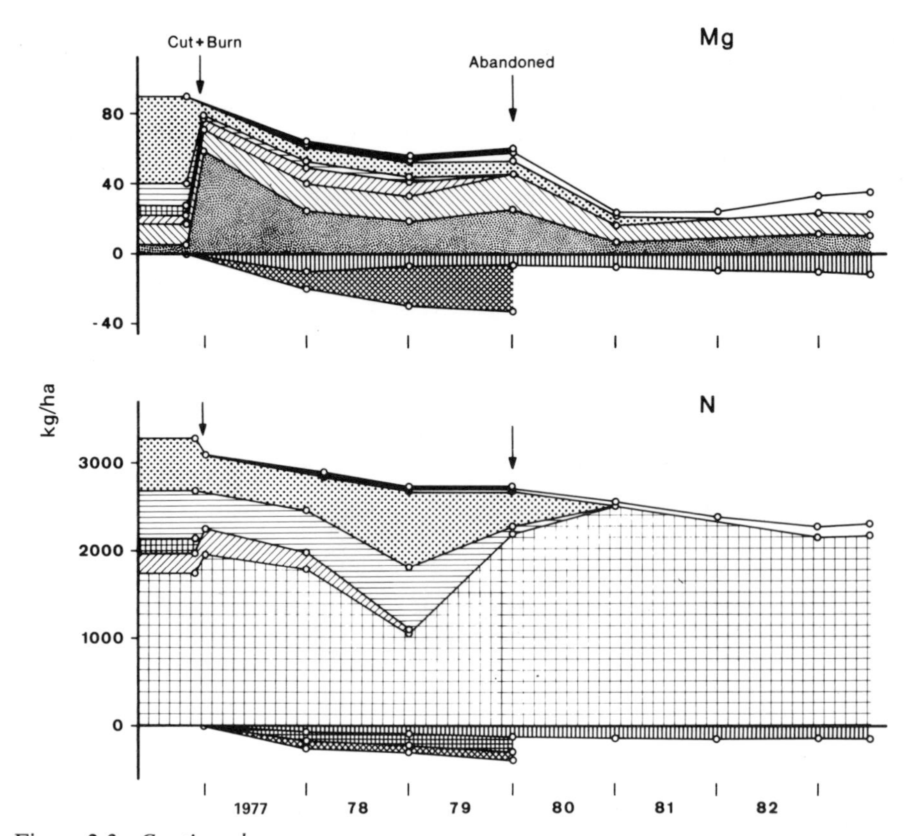

Figure 2.3. *Continued*

harvest and gaseous nitrogen transformations did not continue beyond the point of abandonment of cultivation.

The effect of burning on nutrient stocks in the slash can be seen beneath and to the right of the cut and burn arrow in the graphs. For calcium and magnesium, most of the decrease in stocks in the trunks of primary trees caused by the burn is compensated for by increase in exchangeable stocks in the soil. For nitrogen, there apparently is some volatilization loss during the burn, and the increase in soil nitrogen following the burn is small. The increase in soil potassium following the burn is quickly diminished by leaching losses.

Phosphorus is not shown because, throughout the experiment, most of the phosphorus in the ecosystem remained immobile and fixed in the soil. Of the total 300 kg/ha of phosphorus in the precut forest ecosystem, 251 kg/ha, or 84%, was in the soil. Of that 251 kg/ha, about 2%, or 5 kg/ha, was in soluble or exchangeable form, and the rest was unavailable to crop plants. Following a burn, soluble phosphorus in Oxisols near San Carlos increases by about 10 kg/ha, but this is less than 4% of the total in the ecosystem.

Nutrient Dynamics in Soils

Despite leaching losses, soil stocks of calcium and potassium, and to some extent magnesium and nitrogen, remain fairly constant throughout the period of cultivation. This is because losses from the soil are compensated for by nutrients moving down into the soil from decomposing organic matter on top of or near the surface of the soil. As long as there is a supply of decomposing organic matter on the soil surface, total stocks of nutrients in the soil do not decline greatly. Only after most of the decomposing organic matter on the soil surface has disappeared does there begin a marked decrease of nutrients in the soil. The disappearance of above-ground organic matter is illustrated in Figure 2.3 by the compartments labeled organic matter in root mat and roots and trunks of primary forest. The litter and humus on the soil surface disappeared somewhere between the second and third year of cultivation, and the remains of the primary forest trunks and roots within a few years after that.

Lack of evidence for a sharp decrease in soil nutrient stocks during slash and burn agriculture contradicts the view that soils are depleted of nutrients during shifting cultivation, and that this depletion is an important reason for the rapid decline in productivity usually observed in shifting cultivation (Watters 1971). However, a review of recent literature suggests that the San Carlos finding may not be an exception. There have been a number of studies that have shown little or no nutrient loss during slash and burn agriculture. Nye and Greenland (1964) in Ghana, Zinke et al. (1978) in Thailand, Denevan (1971) in Peru, Harris (1971) in Venezuela, Brinkmann and Nascimento (1973) in Brazil, and Sanchez et al. (1983a) all found that during cultivation levels of most nutrients in the soil did not decline to levels below those of the preburn forest.

It has also been shown for temperate forest ecosystems that when forest disturbances are relatively short, on the order of a few years or less, and when slash is left on the site, declines in soil stocks of nutrients are not as large as expected. In an experimental watershed in West Virginia, Aubertin and Patric (1974) found that there were negligible nutrient losses following clear-cutting carried out so as to resemble conventional clear-cut logging techniques. Studies in the Hubbard Brook region of New Hampshire showed that when herbicides are not used following clear-cutting, nutrient losses are quite low (Pierce et al. 1972). Other studies showed that nutrient losses from leaching and erosion following clear cutting were often short term and small compared to tree removal during harvest (Swank and Waide 1980, Hornbeck and Kropelin 1982, Sollins and McCorison 1981, Kimmins 1977, Cole and Bigger undated). It appears that in temperate as well as tropical ecosystems, small, short-term disturbances do not deplete soil stocks of nutrients to the point where productive potential of the site is seriously inhibited.

Nutrient Dynamics Following Abandonment

Following abandonment of cultivation, there was a sharp drop in nutrients in the soil (Fig. 2.3). The drop could not be explained by leaching losses. Observations of the root mat and mineral soil in the cultivated plot suggested that surface erosion may have caused this loss. The mat of humus and roots on the surface of the soil had scarcely been damaged by the fire, 3 years earlier. It had remained as almost a complete ground cover through the first 2 years of cultivation, although it gradually became thinner. During this time, there was no visible evidence of soil erosion. Only after patches of the mat completely disappeared in 1979, and mineral soil was exposed to the direct impact of rainfall, did observable sheet erosion begin. Where the soil was protected from rainfall, as beneath flat stones, several centimeters of erosion were sometimes evident. Probably not all the eroded soil left the plot. Accumulations of soil occurred behind woody debris, which acted as dams on the soil surface.

The erosion appeared to last only about 1 year. By 1981, the sharp decrease of soil calcium, potassium, and magnesium that had occurred in 1980 had decreased or reversed. A year later, the decline of nitrogen seems to have been halted. The decrease in loss rates may result from an increase in cover over the mineral soil. By 1981, roots and litter from the secondary vegetation invading the abandoned *conuco* almost completely covered the soil surface, and there was little further visible evidence of erosion.

However, stocks of calcium in the soil continued to decline at a low rate, suggesting that the increase in vegetative stocks of calcium came at the expense of soil stocks. Inputs from atmospheric deposition or nitrogen fixation appear to be sufficient to build the vegetative stocks of the other nutrients, because the soil stocks also may start to increase or at least do not decline any further after 1981 or 1982.

Productivity

During Cultivation

Although the nutrient stocks in the soil of the experimental slash and burn plot did not decline greatly during the period of cultivation, total net primary productivity of the crop vegetation declined noticeably, and total yuca root harvest was reduced to half (Fig. 2.4). Total net primary productivity of crop vegetation dropped from 5.3 to 3.1 t/ha/yr, and dry weight of edible yuca production dropped from 1.46 to 0.70 t/ha/yr between the first and third years.

Total net primary productivity of the control forest changed very little throughout the course of the experiment (Fig. 2.4), suggesting that changes in crop and weed productivity did not result from such extrinsic variables as climate.

Reasons for Declining Productivity

The discussion above suggested that changes in the total stocks of nutrients in the soil were not likely to be an important factor causing the decline in crop productivity. However, the proportion of the total nutrient stocks in the soil that is available to the crop plants is another factor, not easily determined. At the time this project began in the early 1970s, the importance of distinguishing

Figure 2.4. Net primary productivity, in tons dry weight per hectare per year of total crop plant biomass, edible crops (tubers of yuca), and secondary successional vegetation (weeds) in the experimental plot, and total wood, roots, and leaves of trees in the control plot.

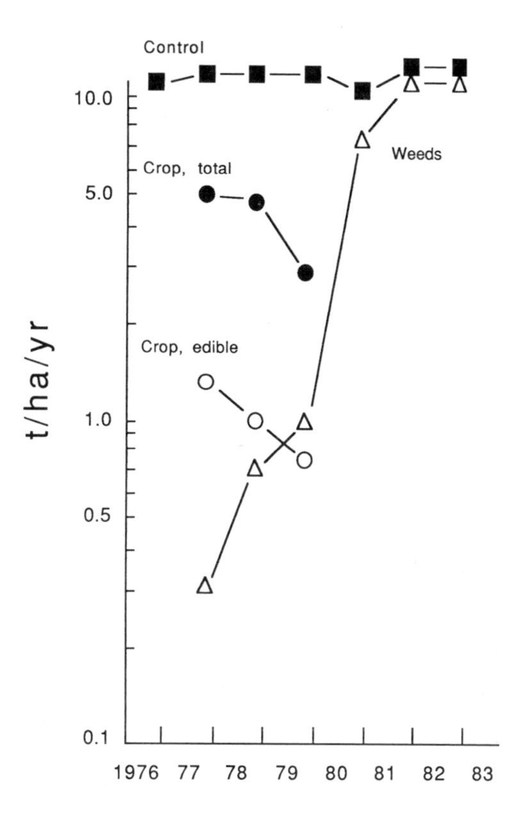

between the various forms of soil nutrients, such as between labile, organically bound, and sorbed phosphorus, had not been appreciated fully. Initially, only total soil phosphorus was measured following a total Kjeldahl digestion. Only later did we realize that a change in the form of soil phosphorus from a more to a less soluble form could be important in the decline in crop productivity.

Although throughout the entire experiment only total soil phosphorus was measured, soil pH was measured, and availability of phosphorus in acid Oxisols often is correlated with soil pH (Sanchez 1976). Soil pH went from 3.9 in the preburn forest, to 5.4 after the burn, to 4.1 at the time of abandonment of cultivation in 1979, to 3.8 in 1983 (Fig. 2.5). A soil study (C. Uhl, unpublished data) of slash and burn sites of various ages in 1984 showed that at pH of 4.2, available phosphorus in Oxisols near San Carlos is about 4.8 ppm and increases to 13 ppm as soil pH reaches 5.1. This correlation of available phosphorus with soil pH, and the pattern of declining pH in the experimental site during cultivation, suggests that availability of phosphorus in the soil may have been an important factor responsible for declining crop production during the period of cultivation. Decreasing pH also results in increasing availability of aluminum in the soil (Sanchez 1976), and aluminum toxicity also may be a factor in declining productivity of crop plants.

Sucking insects removed up to 14% of the radioactive phosphorus injected into the stems of crop plants (Montagnini and Jordan 1983). However, the

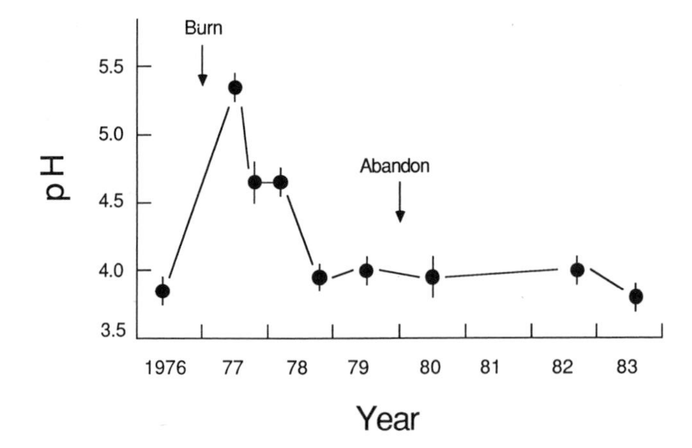

Figure 2.5. Soil pH and standard deviations in experimental slash and burn plot, beginning with undisturbed forest and extending through cut and burn, cultivation, and abandonment.

insect species seemed to be restricted to the *conuco* area and to return the nutrients to the soil within the *conuco* site. Chewing and boring insects consumed less than 3% of plant tissue, possibly because of the presence of cyanogenic glucosides in the cassava plants (Montagnini and Jordan 1983). Herbivory probably was not an important cause of the decline in crop productivity.

Although the increase in weed (successional vegetation) productivity between the first and third year of cultivation (Fig. 2.4) suggested that weed competition might have played a role, weed productivity at the time of abandonment was only about one-third that of total crop plant productivity. Weeds never appeared to overtop or crowd out the crop plants.

Productivity Following Abandonment

After weeding ceased following abandonment of cultivation, productivity of successional vegetation increased rapidly. By the second year following abandonment productivity was about equal to that of the undisturbed control forest (Fig. 2.4). The much greater competitive ability of the successional vegetation in the abandoned *conuco* suggests that the native vegetation can acquire nutrients, particularly phosphorus, that are relatively unavailable to crop plants. The literature on adaptations of native plants that enable them to be better competitors than crop plants has been reviewed by Jordan (1985a). The following list is a summary of these adaptations:

1. A large root biomass, which allows a more complete exploitation of the soil
2. A high root/shoot ratio, which helps provide a greater supply of nutrients for photosynthetic tissue
3. A relatively long life span, which allows plants to take up nutrients during seasons of abundance and store them for times of deficiency

4. Low nutrient-uptake kinetics of the roots, which allows a better adaptation to the low nutrient-supplying power of the soil
5. Ability to grow and survive with low tissue nutrient concentrations
6. Mycorrhizae that are better adapted to local conditions
7. Tolerance of acid soils and high aluminum availability

Conclusion

Indigenous slash and burn agriculture resuits in a loss of nutrients from the soil. However, losses from the soil are compensated for by input from decomposing slash and other organic material on the soil surface. Soil stocks remain relatively high, therefore, while total ecosystem stocks decrease during the period of cultivation.

Because soil stocks remain relatively high, it is unlikely that low nutrient stocks are responsible for the sharp decline in crop productivity. Instead, it may be the availability of nutrients, especially phosphorus, to crop plants that limits their productivity.

In contrast to crop plants, native successional species invading an old *conuco* do not appear to suffer from low availability of nutrients. The wild species are adapted to take up nutrients relatively unavailable to crop species.

Cultivation of Amazonian soils, when practiced in the traditional manner, does not appear to inhibit the beginning of natural secondary succession. Whether succession will continue, and a forest will develop similar to that existing prior to the cutting, is another question. Long-term successional dynamics and the resilience of the forest following shifting cultivation are the subject of the next chapter.

3. Recovery Following Shifting Cultivation* †

Case Study No. 2: A Century of Succession in the Upper Rio Negro

Juan G. Saldarriaga

Does slash and burn agriculture in Amazonia permanently change the rain forest, or can a disturbed area recover fully? Chapter 2 data on the productivity of early successional species invading the abandoned *conuco* suggested the site was well on its way to recovery. Other evidence, however, raises doubts about potential resilience of the forest. Total stocks of calcium, potassium, and magnesium in the abandoned *conuco* were considerably below the levels needed to rebuild a mature forest. To answer questions about recoverability of the forest following shifting cultivation and about rates at which nutrients are restocked, this study examined ecosystem dynamics over long-term succession in the upper Rio Negro.

Long-term forest succession cannot be studied by monitoring one stand or a few stands over the entire time span of a successional sere. The process is too slow. A more practical approach is to study a chronosequence of sites that appear to have been similar before disturbance and that differ only in length of time since abandonment of cultivation. This approach was used to study recov-

---

* Except where otherwise cited, descriptions, data, and findings in this chapter are from: Saldarriaga, J. G., 1985. *Forest Succession in the Upper Rio Negro of Colombia and Venezuela*. Ph.D. dissertation, Ecology Program, University of Tennessee, Knoxville, Tennessee, USA.

† Publ. 2660, Environmental Sciences Div., ORNL.

ery from slash and burn agriculture in the San Carlos region of the upper Rio Negro.

Methods

Site Selection

The study was conducted near San Carlos de Rio Negro, Venezuela. To obtain a sufficient number of sites for a chronosequence, an area of about 200 km^2 around San Carlos was surveyed. Within this area, 23 sites were selected for intensive investigation (Fig. 3.1). All sites selected were *tierra firme* sites on Oxisol or Ultisol, soils commonly used for shifting cultivation in the region.

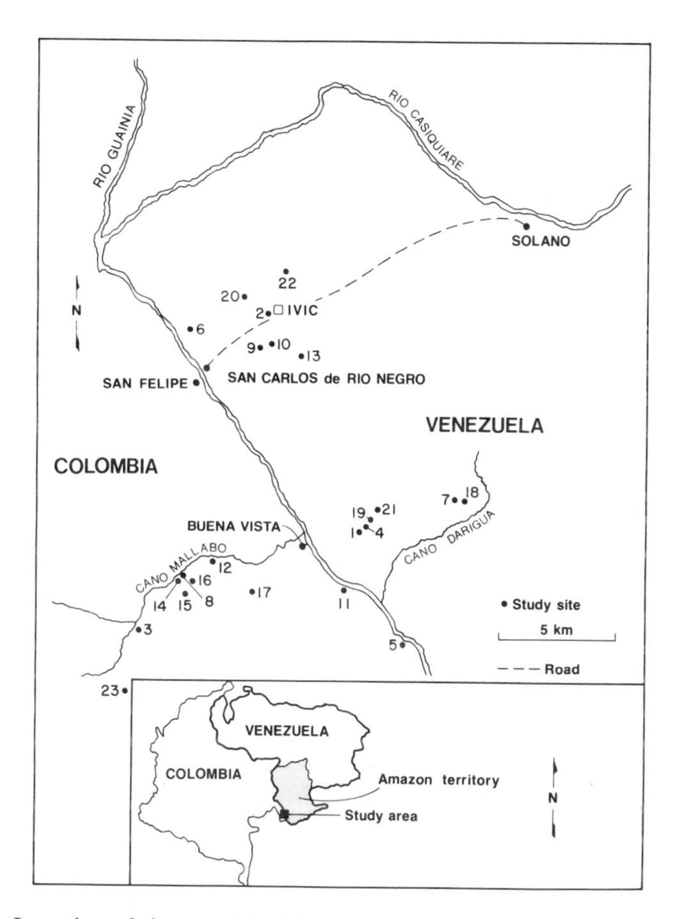

Figure 3.1. Location of sites used in this study, shown in relation to San Carlos de Rio Negro. Each number corresponds to a site. The slash and burn site in Chapter 2 is also included and is labeled IVIC, an acronym for Instituto Venezolano de Investigaciones Cientificas, the laboratory that supported the field site.

None of the sites was on Spodosol. Sites were chosen only where time of abandonment from cultivation could be established reliably. Sites very close to towns or main paths were excluded so that human disturbance since abandonment would not be a factor.

Sites under 30 years old were dated by farmers living in San Carlos who could make a reliable estimate of time of abandonment. For fallows older than 30 years, age was established by interviewing long-time residents of the region and correlating their accounts of when they lived at the sites with historic events that had well-established dates. Some elderly people interviewed were children when the rubber boom ended around 1910. During the boom, almost all adult Indian males in the San Carlos area collected latex from wild rubber trees. When the boom ended, many families were affected. Another important event in the upper Rio Negro was the political movement organized by Tomas Funes. Between 1913 and 1921 he led a terroristic dictatorial regional government in the Venezuelan Amazon Territory. During this time many villages and towns were abandoned, and their inhabitants fled into the forest. Many of the old stands, called *rastrojos*, found today near San Carlos are results of activities at this time. The impact of these events on the lives of these people was significant enough so that they could remember where they were when these events occurred.

Sites were classified into six age categories: 9–13 years (N = 4); 20 years (N = 4); 30–40 years (N = 4); 60 years (N = 3); 80 years (N = 4); and mature *tierra firme* forest (N = 4). Sites of mature *tierra firme* forest were those which long-time residents of the region believed had never been cultivated.

Biomass Studies

At each site, three 10 × 30 m plots were established. In all plots, individuals with a diameter at breast height (dbh) of more than 1 cm and a height of more than 2 m were identified by local name, and dbh was measured. In addition, three 152-m^2 subplots were used for trees with diameters of less than 1 cm and heights of 1 to 2 m. Both living and dead trees were measured.

To calculate biomass as a function of stem dbh, height, and wood density, 126 trees were harvested and measured, and their dry-weight biomass was determined. Twenty-eight trees were from mature forest or 80-year-old forest. Sixteen were from fallows 10–35 years old. In addition, 18 palms, 10 *Heliconia* sp., and 54 small trees (1–10 cm dbh) were taken from fallows or mature forest. Biomass data collected by Jordan and Uhl (1978) were added to these data to increase the number of species and trees larger than 40 cm dbh. Regressions were developed and used to predict biomass of all trees in the study plots, and individual values were summed to give above-ground stand biomass.

Root biomass was determined from four pits (50 × 50 × 100 cm) at each of six of the study sites, representing the full range of age classes. Soil was removed from the pits in 10-cm layers, and the roots were separated from the soil by hand, cleaned, weighed, and dried to constant weight. Root biomass was correlated with site basal area, and root biomass at sites without pits was estimated by these correlations.

Nutrient Stocks

At each of the 23 study sites, three 32-m transects were run and soils on each transect were sampled at 1-m intervals. Samples from a 0–15 cm depth were taken every meter, and from a 15–30 cm depth every 2 m. Samples were pooled systematically to obtain 12 samples per site from 0 to 30 cm. Total and exchangeable nutrients were determined by standard procedures (Black 1965).

To determine total soil stocks, the concentration of nutrients in the soil must be multipled by the mass of the soil to the depth of rooting. By carefully excavating and weighing the soil from 18 0.25-m^2 soil pits, Stark and Spratt (1977) found that the average weight of the soil to a depth that included 89% of the fine root biomass was 1381 t/ha. This soil mass value was applied to all sites in this study.

Nutrient concentrations in leaves, stems, and roots of the four earliest successional sites were taken to be the same as concentrations in the 5-year-old successional vegetation at San Carlos reported by Uhl and Jordan (1984). Concentrations in the mature forest were taken to be the same as those in the primary forest (Jordan 1980). Concentrations for intermediate stands were assumed to be intermediate, and values were assigned by interpolation. Total nutrient stocks were calculated by multiplying concentrations for the vegetation components at each site by the biomass at each site.

Results and Discussion

Diversity

Diversity of the various age-class sites is illustrated in Figure 3.2. The number of species is plotted as a function of the number of equal-area stands sampled. As often occurs in tropical rain forest sites, the species area curve gives no indication of leveling off as the area sampled increases. Nevertheless, the sites can be compared because all sampling sites have the same area. Clearly, species diversity increases with increasing age of fallow. Especially marked is the increase between the youngest stands and those of 60–80 years.

For trees of 10-cm dbh or larger, the 80-year-old stands averaged 23.5 species per plot and the mature forest averaged 23.7 species per plot. For trees of 1-cm dbh or larger, the 80-year-old stands averaged 71 species per plot and the mature forest averaged 75 species per plot. Diversity data therefore suggest that the 80-year-old stands have almost completely recovered to the value of the mature forest.

However, when the type of species is considered, it becomes evident that recovery to primary forest is still far from complete. Of the 22 most important

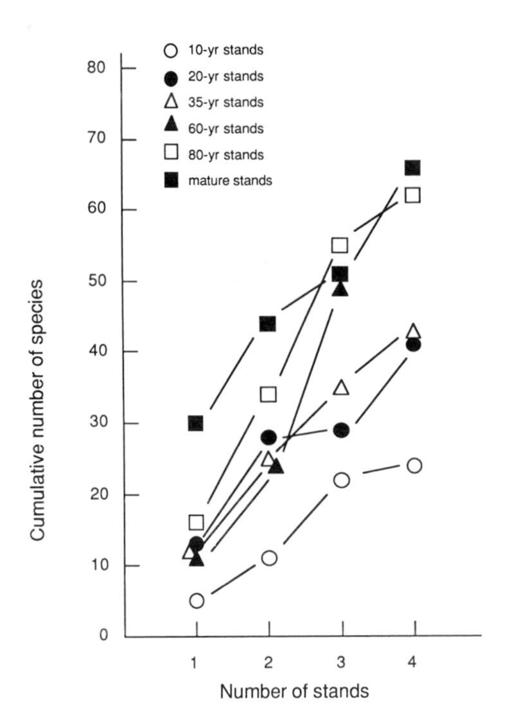

Figure 3.2. Cumulative number of species larger than 10-cm dbh in stands of similar age. All stands were sampled with three 300-m^2 plots.

species as determined by importance values in the 80-year-old and the mature forest stands, only seven species were found in both communities. All other species were exclusive to one or the other.

Structure

A structural change in communities as a function of stand age is reflected in changing density of the wood (Fig. 3.3). These data suggest three distinct types of communities: early successional up to 20 years, midsuccessional through 80 years, and mature forest.

Changes in basal area and forest biomass as a function of stand age are shown in Figures 3.4 and 3.5. Values for recently abandoned sites (age = 0) are from Uhl (1982). Basal area and forest biomass increase rapidly during the early stages, but between 60 and 80 years little change occurs. A linear extrapolation (the dashed lines in the figures) might suggest that basal area and biomass in areas disturbed by shifting cultivation would never recover to values of the mature forest. However, if the 80-year old stand is only midsuccessional, as suggested by species composition and wood density data, invasion of stands by primary species may result in a resurgence of biomass and basal area increment. This would result in a pattern of complete recovery suggested by the solid lines in Figures 3.4 and 3.5.

In contrast to biomass and basal area, leaf area index increases rapidly and reaches a maximum value at about 20 years (Fig. 3.6). Despite changing species

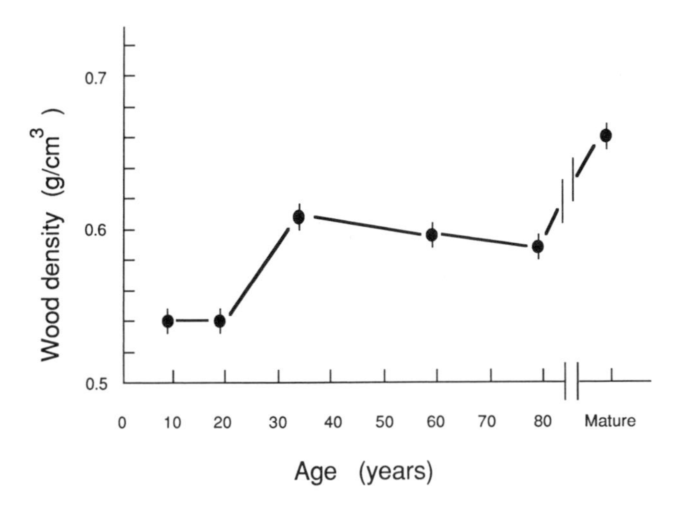

Figure 3.3. Average wood density (dry-weight specific gravity) and standard errors for species in various age classes near San Carlos.

composition after that time, the leaf area index changes little throughout the course of succession through mature forest.

Nutrient Accumulation

The pattern of accumulation of potassium (Fig. 3.7) in the vegetation reflects the pattern of biomass accumulation. A rapid initial accumulation is followed by a leveling off and even a slight decrease at 60–80 years. After that, accumulation would continue if recovery continued toward mature forest.

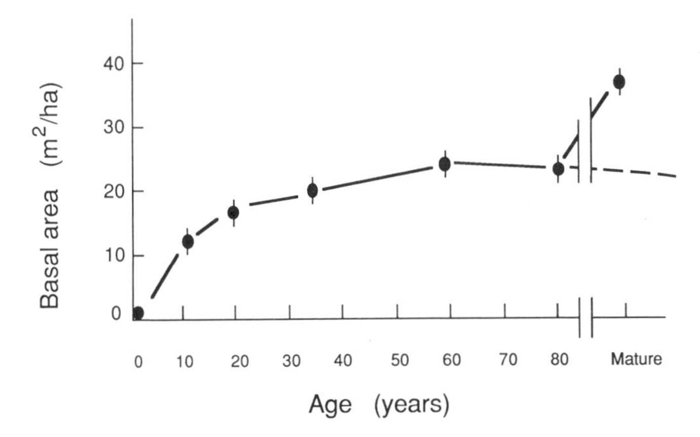

Figure 3.4. Average basal area and standard errors for forest sites of various age classes near San Carlos.

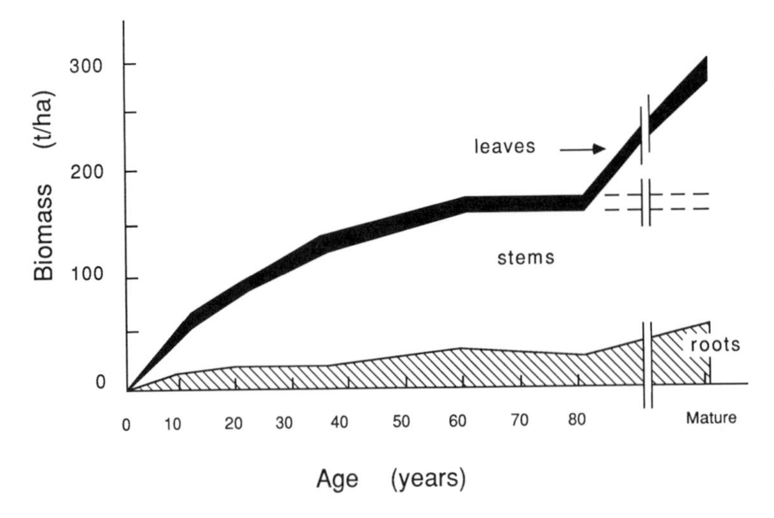

Figure 3.5. Total stocks of biomass as a function of age in stands of the chronosequence. Totals are the sums of stocks in roots, stems, and leaves. Individual components are stacked on top of each other. The stock of each component is indicated by the thickness of the respective shaded, clear, or hatched areas.

Potassium stocks in the soil were similar in all stands of the chronosequence (Fig. 3.7). The soil is chiefly quartz, kaolinite, and iron and aluminum oxides, and weathering of these materials will not be an important new source of potassium for the trees. The increase in potassium stocks in the biomass probably came from greater wet and dry atmospheric deposition than leaching from the soil.

Stocks of exchangeable phosphorus in the soil were similar in all stands of the chronosequence (Fig. 3.8). Exchangeable phosphorus is only a small percentage of total soil phosphorus (Chapter 2). The increase in phosphorus in the

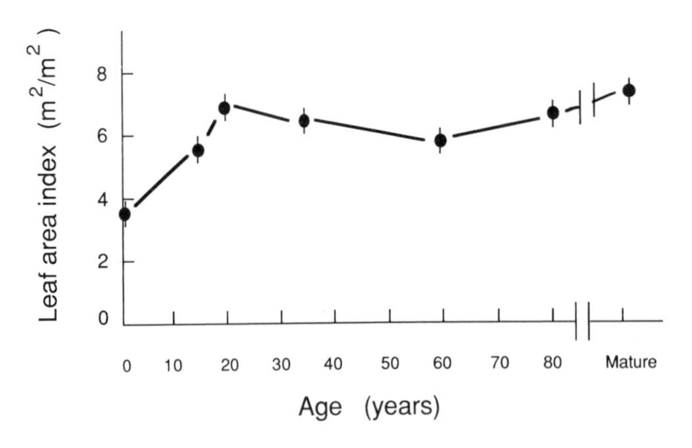

Figure 3.6. Average leaf area index and standard errors for forest sites of various age classes near San Carlos.

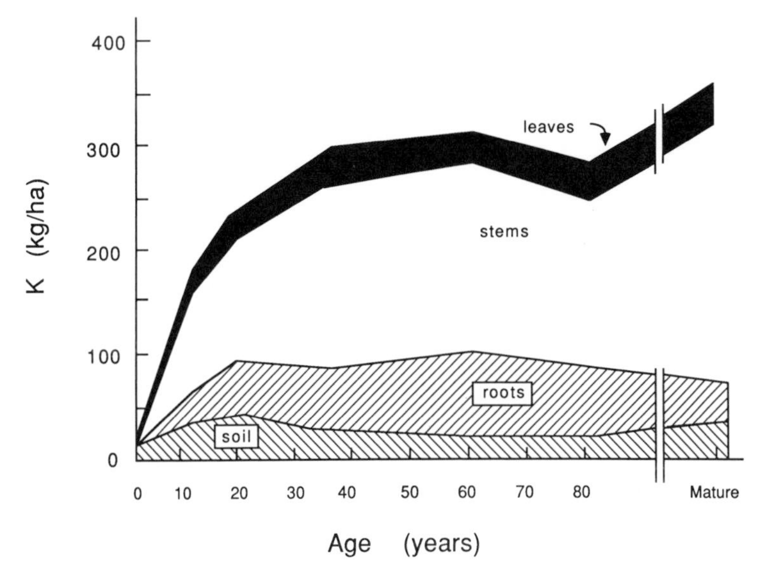

Figure 3.7. Total stocks of potassium as a function of age in stands of the chronosequence. Totals are the sums of stocks in soil, roots, stems, and leaves. Individual components are stacked on top of each other. Stock of each component is indicated by the thickness of the respective shaded, clear, or hatched areas. Nutrient concentrations used for calculation of soil stocks are not included in Saldarriaga's 1985 dissertation but are available from the author.

accumulating biomass probably came from phosphorus bound in relatively insoluble form in the soil. Native species apparently are relatively effective at taking up this bound phosphorus (Chapter 2).

Calcium stocks in the biomass (Fig. 3.9) are probably built up primarily as a result of accumulation of atmospheric deposition. During the first 35 years, buildup may occur in part from soil stocks, as evidenced by their slight deple-

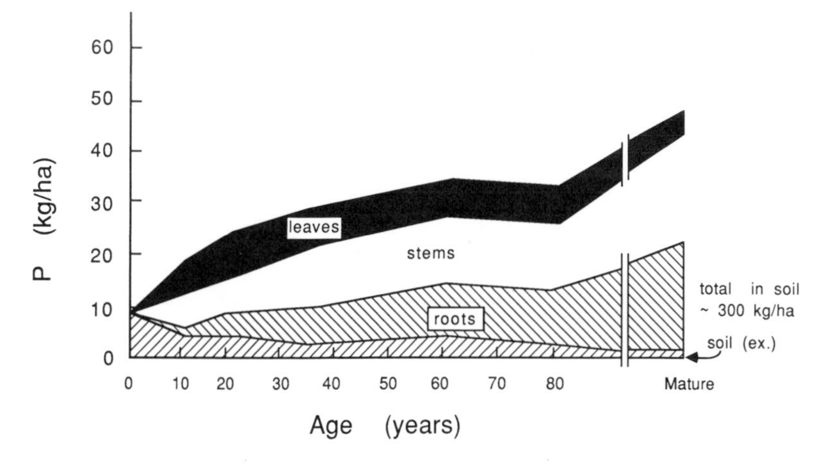

Figure 3.8. Phosphorus stocks. Conventions as in Figure 3.7.

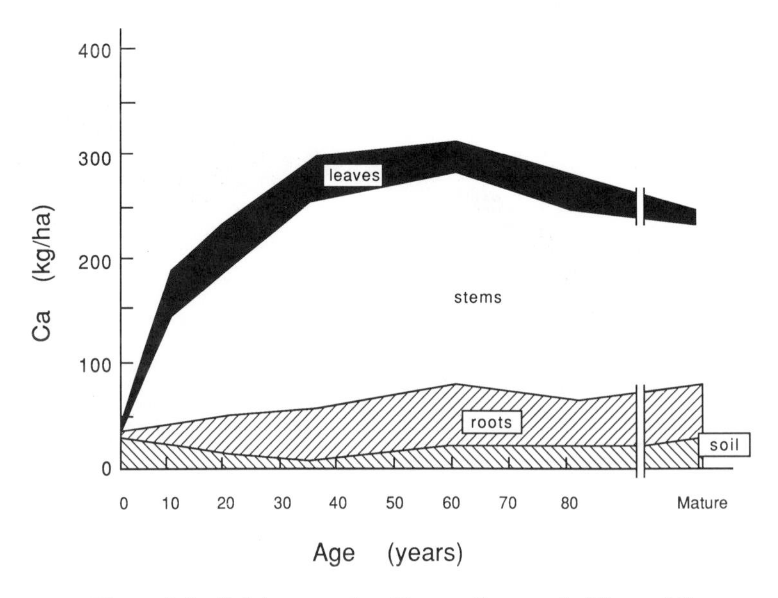

Figure 3.9. Calcium stocks. Conventions as in Figure 3.7.

tion during this time. Total ecosystem stocks of calcium in the chronosequence show a peak at 60 years. After that, a decline follows through to the mature forest. The decline results from much lower concentrations of calcium in the tissue of mature forest species than in tissue of the early successional species. In early successional species leaf calcium averages 5.12 mg/g; in mature forest leaves calcium concentration averages 1.33 mg/g. In early successional species

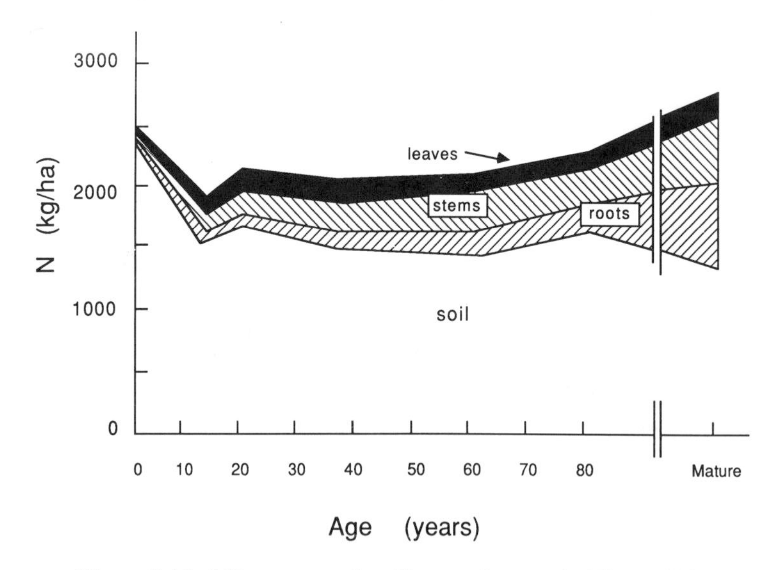

Figure 3.10. Nitrogen stocks. Conventions as in Figure 3.7.

wood calcium averages 2.01 mg/g; in mature forest wood the average is 0.73 mg/g.

In contrast to the other nutrients, total ecosystem stocks of nitrogen continued to decline for another 10 years after abandonment of cultivation (Fig. 3.10). Nitrogen in living biomass increased slightly early in the chronosequence. The sharp drop was in soil stocks, probably because nitrogen loss from leaching and volatilization occurred at rates greater than atmospheric deposition and fixation. Later in the chronosequence nitrogen builds up gradually toward mature forest levels, without the decline shown by potassium and phosphorus.

Conclusion

If the chronosequence were a single successional sere, the data suggest that it takes about 80 years for the biomass to build up to a value of about half of that of the mature forest. Would a site that was abandoned and undisturbed for one to two centuries eventually return to primary forest?

Because this study includes no sites between 80 years and mature forest, it lacks conclusive evidence that sites of shifting cultivation eventually redevelop the structure of a mature forest. However, the trends established during the first 80 years suggest that a return to mature primary forest is possible. Total ecosystem stocks of calcium and potassium increased apparently because of atmospheric input, and this process can continue. Phosphorus in biomass increased, probably through mobilization of insoluble reserves in the soil. Nitrogen increased, probably in part because of biological nitrogen fixation. Species composition in the 80-year plots seems to be shifting toward that of a primary forest community.

Abandoned sites of shifting cultivation in the upper Rio Negro region will probably eventually return to mature forest. However, the process may take more than a century. Cultivation after a fallow of less than a century is possible, but nutrient stocks will not have been completely rebuilt. Nutrients stocks would be less than the those at the beginning of the cultivation following cutting and burning of primary forest, as described in Chapter 2.

4. Shifting Cultivation Where Land Is Limited*

Case Study No. 3: Campa Indian Agriculture in the Gran Pajonal of Peru

Geoffrey A.J. Scott

Introduction

Shifting cultivation in the Amazon Territory of Venezuela (case study No. 1) generally takes place following cutting of primary forest. Because population density is very low there is little pressure to cut secondary forests. Therefore, long fallows are possible, during which nutrient stocks appear to be replenished to the level of the predisturbance forest (case study No. 2).

In contrast to the situation near San Carlos Venezuela, the shifting cultivation described in this case study takes place where available or convenient arable land is limited. Forest fallows are cut before primary vegetation becomes reestablished. Sometimes relatively young secondary vegetation is cleared to establish plots for agriculture. The frequent occurrence of fire in abandoned sites of cultivation also strongly influences the successional vegetation.

Site Description

The Amazonian sector of Peru is divided into the low selva (*selva baja*) below 400 m, and the high selva (*selva alta*) between 400 and 2000 m. Within the high

---

* Except where otherwise cited, descriptions, data and findings in this chapter are from: Scott, G. A. 1979. *Grassland Development in the Gran Pajonal of Eastern Peru*. Ph.D. dissertation, Department of Geography, University of Hawaii at Manoa, Honolulu, Hawaii, USA.

selva there is a region of approximately 3500 km^2 located north of the Tambo and west of the Ucayali rivers (Fig. 4.1) called the Gran Pajonal. It is an uplifted block–plateau of Permo-carboniferous shales, limestones and sandstones, averaging about 1000 m above sea level, deeply dissected by incised streams. The area receives annually between 1700 and 2100 mm of precipitation, most of which falls between October and May.

The vegetation in the Gran Pajonal consists of some primary forest and a mixture of secondary forest, stands of bracken fern, savanna, and grassland, the latter three formations all strongly influenced by fire. When the first missionaries arrived in the region, they were so taken by the short grass savannas amidst tall forests that they named the region the "great grassland" or Gran Pajonal, despite the fact that grassland savanna covers only 3% of the total region. The grassland savannas were produced by Indians during a long history of shifting agriculture and burning.

The Gran Pajonal is inhabited by the Campa Indians, or Ashaninka as they call themselves. They are a proud and independent people and resisted outside interference for more than two centuries. Early attempts by Franciscan missionaries to "subdue and Christianise the infidels" in the 1730s were short lived. Permanent contact with the region was established only in the late 1930s when settlers established the small mission of Obenteni.

Because of migration of Peruvian settlers eastward from coastal or Andean cities, most of the Campa are now within the settlers sphere of influence and economy. The Campa today number about 24,000, but of these only 1800 living in the Gran Pajonal have managed to retain their traditional language and way of life. Some Campa live close to Obenteni and work part time for settlers, but many have moved into more remote regions of the Gran Pajonal.

Campa Agriculture

Sites of Indian huts are usually located in the center of ridgetop savannas, but some are also found in new forest clearings. The hut is surrounded by a bare earth area 20 m in diameter, and rarely are more than 10 people found living together. The Campa have no villages and population densities range from less than one to three per square kilometer.

The mainstay of the Campa is the garden, or *chacra,* in which are grown almost all of their carbohydrate requirements and some protein as well. The principal crop, sweet manioc (*Manihot esculenta*), makes up 75% of their diet and is also used for making the local beer, *masato*. Other important garden crops include plantain, bananas, corn, sweet potatoes, beans, and occasionally peanuts. The achote shrub is also grown to provide seeds, which are ground with animal fat to produce red paint for facial colorings. Tobacco and coca are grown for chewing, while the roots of the barbasco shrub produce a fish poison, and cotton is used for weaving the traditional Campa robe, the *cushma*. Gathering, hunting, and fishing provide meat, snails, insects, and some fruit and mushrooms. Bird plumage is used as a decoration on their ceremonial hats,

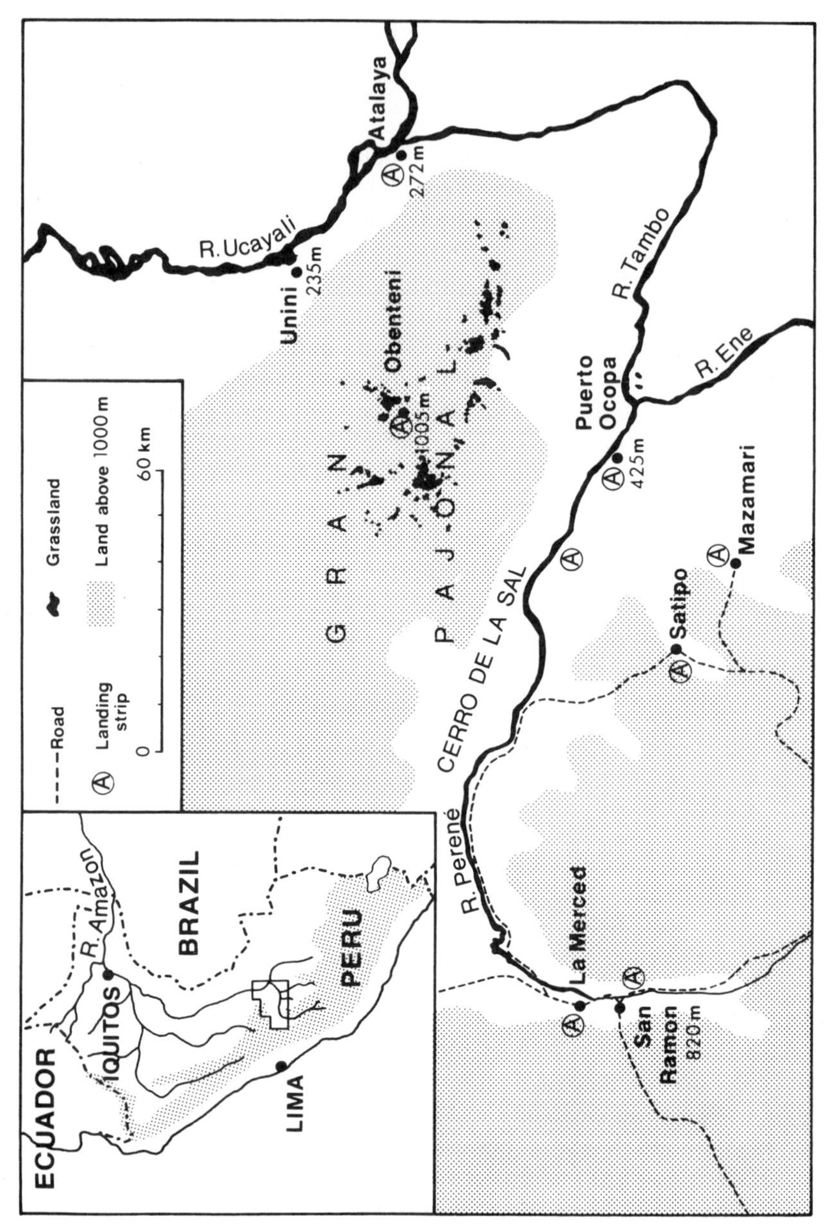

Figure 4.1. Map showing location of Gran Pajonal.

called coronets, as well as on the *cushma* and as stabilizers on the 150-cm-long arrows. Puma, jaguar, monkey, and tigrillo skins are used for mats.

Gardens are prepared at the beginning of the dry season in June. Usually about 1 ha of forest is felled after undergrowth has first been cleared with machetes. Clearing the undergrowth usually takes 3 weeks, whereas cutting the trees requires only a couple of days. The undergrowth must be killed, or it will suddenly emerge in vigorous growth that will shade the trees after they are cut and prevent this slash from drying out. Felled trees, that is, the slash, are allowed to dry out for several months before burning is attempted.

After the burn, the crop species are planted in the soil, now enriched with ash. Although the plots appear extremely disorganized because of the tree stumps and unburned trunks, the gardens are in fact carefully designed to maximize production, minimize work, and protect the soil. Often 40 planted species can be found in the *chacras* of one Campa family. Because tubers of manioc can take almost 1.5 years to reach maximum size in this region, a Campa family will usually have two or three gardens of different ages going at the same time. Often in as little as 3 years, declining productivity causes the garden to be abandoned and the Campa farmer shifts his activities to another area, where forest is felled and a new garden is initiated.

When gardens are abandoned they quickly revert to secondary forest. In 15 years secondary growth can reach a height of 15 m, and sites may then be cut and cultivated again, or succession may be allowed to proceed. If the Campa do not want to move when a garden is abandoned, but instead cut new gardens farther downslope, they can prevent the newly abandoned garden from developing into thick impenetrable secondary forest by periodic burning. If the abandoned garden is burned annually during the dry season, a fire subclimax develops. Whereas most fires are set intentionally for clearing or maintenance purposes, some are set purely for fun, or to provide a "window to the sky."

Grassland Development

The evidence from phytosociological studies in the Gran Pajonal suggests the following sequence of vegetation change resulting from fire-associated activities of the Campa.

1. Primary or secondary forest is cut, burned, and planted to crops.
2. The cultivated field (swidden, or *chacra,* which is the local term) is used from 1 to 4 years. The dominant crop is *Manihot esculenta* (manioc, yuca, or cassava). Dominant weeds are *Pteridium aquilinum* (bracken fern, or *chac-chac*) and *Imperata brasiliensis* (a coarse grass), and a woody species, *Baccharis floribunda*.
3. Following abandonment from cultivation and initiation of annual burning, the *chac-chac* stage dominated by *Pteridium aquilinum* begins. Some erosion of the soil surface is evident.
4. *Chac-chac* gives way to a cover dominated by *Imperata brasiliensis*. Soil erosion continues, infiltration rates decrease, and there is some invasion by *Andropogon* spp. (*Andropogon* is a genus of grass, many species of which

are found on soils low in nutrients, and sometimes called "poverty grasses.")
5. Youthful grassland is dominated by several species of *Andropogon,* and some *Imperata* is still present. Erosion and deterioration of soil continues.
6. Old grassland is almost completely dominated by *Andropogon* spp. and small sedges.

In the absence of burning, secondary succession on the abandoned *chacra* plots consists mainly of woody species. Studies of the accumulation of biomass and nutrients during secondary succession, as well as during formation of old grassland, were the major focus of the research described here.

Methods

Biomass and Age of Vegetative Communities

The study of vegetative change from abandoned swidden back to forest, and from abandoned garden to fern and grassland as a consequence of burning, included various approaches. The data on stand age came from aerial photographs from 1953 and 1958 and from personal observations in 1969, 1971, and 1976. Interviews of natives and colonists also helped establish the ages of certain communities.

Data on the biomass of herb, grass, and fern communities came from transect and quadrat studies in conjunction with allometric relationships derived from selected harvests. The number of quadrats and trees sampled varied depending on the vegetation type. In each quadrat, cover and other measures of dominance were determined (Mueller-Dombois and Ellenberg 1974, Braun-Blanquet 1965, Cain and de Oliveira Castro 1971). Selected quadrats were completely cropped, all roots extracted, and oven-dry weights determined.

In secondary and primary forest communities, square quadrats with 30-m sides were established. In each quadrat, cover, height, and diameter at breast height (dbh) were measured for all trees. Species were differentiated by Campa Indian guides and given local names. An allometric relationship between diameter at breast height and oven-dry weight of sample trees was determined following the procedure described by Ogawa et al. (1961). Data on number of individuals in each size class per hectare, and average weight of trees in each size class were used to predict biomass per hectare. Soil samples were also collected in each plot and analyzed for organic matter content and bulk density.

Nutrient Stocks

Samples of roots, stems, and leaves were collected for dominant tree species, and composite vegetative samples were taken from swidden, fern, and grassland communities. Soil samples for chemical analysis were taken in each community type at 6-cm depth intervals to 42 cm, and also by soil horizon. Chemical analyses were carried out at La Estacion Experimental Agricola de La Molina, Lima, Peru. Soil bulk density was also determined. Total nutrient stocks in soil and vegetation components was a product of nutrient concentration times mass of the respective components.

Results and Discussion

Biomass Dynamics

Forest felling and burning of slash cause immediate changes to the biomass and nutrient stores of the preexisting forest community (Fig. 4.2). Although much depends on the age of forest stands cleared, the most immediate changes are in the leaf, branch, ground flora, and litter components. When mature forest is felled, there are trunks and stumps as thick as 1 m or more. These do not burn but remain to decompose slowly within the swidden. In the case of young secondary forest, however, where diameters often do not exceed 15–20 cm, practically all slash is consumed. Figure 4.2 indicates a large loss of trunk biomass during burning, but this is because the preburn values are of primary forest and postburn values are from sites derived from secondary forest.

During the 2- to 4-year life of a *chacra,* biomass levels in dead trunks, stumps, and litter decline. At times, between weedings, weed biomass increases occur. Crop biomass also fluctuates, depending on growth and harvesting rates. If a *chacra* is close to the Campa hut, logs also will be removed frequently for household needs. Soil profile studies showed that litter horizons practically disapper during the life of a *chacra,* a change attributable to water erosion, wind removal, decomposition, and incorporation into the profile by soil fauna. By the time *chacras* are abandoned, most trunks and stumps have decomposed partially.

The 3-year-old stand of secondary forest had a living biomass of 52 t/ha dry weight and a total biomass including litter, of 65 t/ha. A 10-year-old stand at Gran Pajonal dominated by *Cecropia* had a live biomass of 89 t/ha and a total biomass of 103 t/ha. Total biomass in a stand between 25 and 30 years old was 185 t/ha. Biomass in areas of the Gran Pajonal that are subjected to frequent burning following abandonment of cultivation have much lower stocks of biomass (Fig. 4.3).

The biomass values for 30-year-old unburned successional forests in the Gran Pajonal are higher than values for comparably aged stands at San Carlos (Fig. 3.5, Chapter 3). The greater biomass and higher productivity at the Gran Pajonal sites compared to San Carlos may result from more fertile soils at the Gran Pajonal (see next section).

Nutrients

Comparisons of Figures 4.4–4.8 and Figure 2.4 show that stocks of nitrogen, potassium, calcium, and magnesium are all much higher in the vegetation and soils of the primary forest at Gran Pajonal than at San Carlos, Venezuela. There are large differences in both the soil and the vegetation. Phosphorus stocks also were higher in the primary forest vegetation of the Gran Pajonal, where there were 160 kg/ha compared to about 50 at San Carlos (Chapter 2). Exchangeable phosphorus at both sites was low, between 5 and 10 kg/ha, but exchangeable phosphorus probably is not a good index of phosphorus available to native species because of adaptations that allow them to take up phosphorus that is relatively insoluble (Chapter 2).

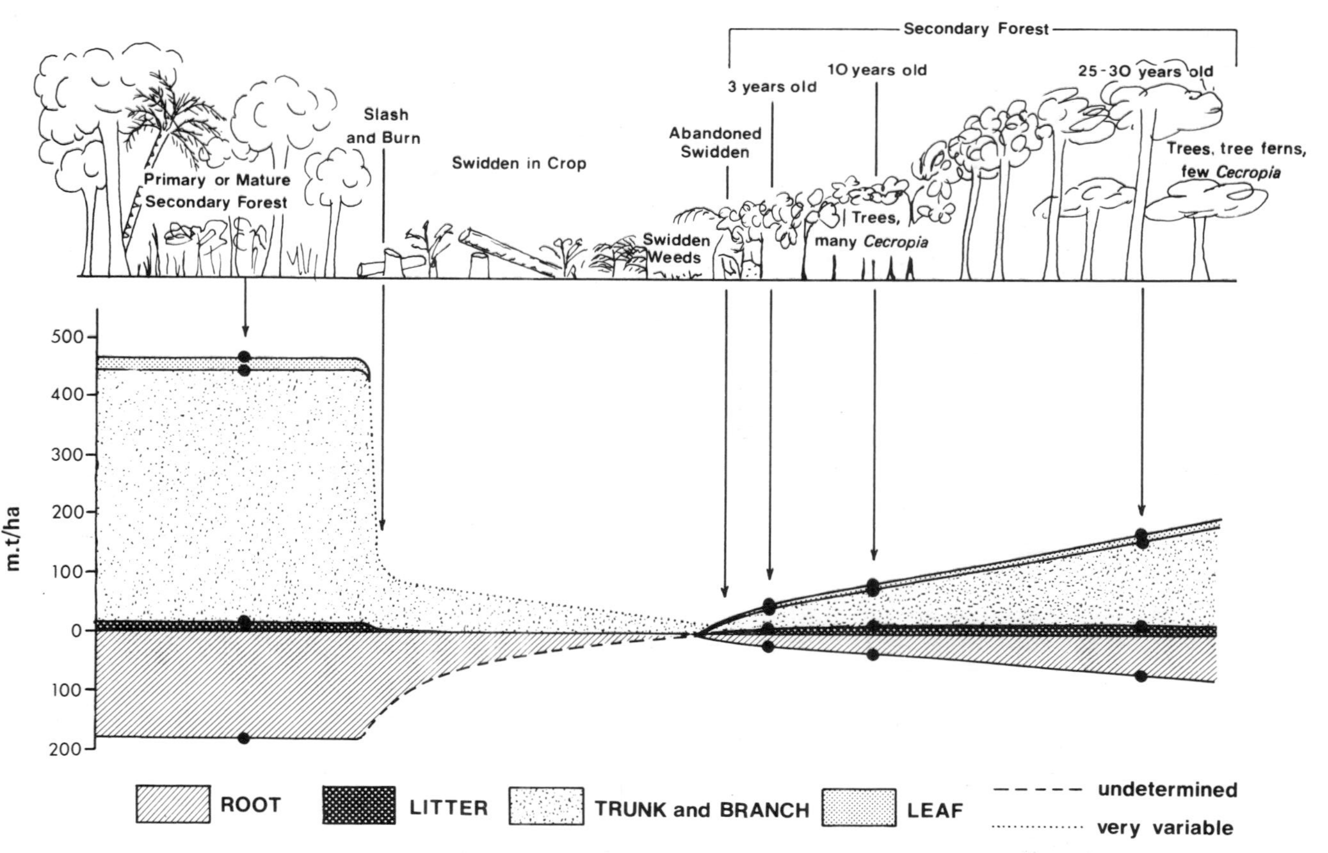

Figure 4.2. Changes in biomass stocks as mature forest is cleared for swidden and then allowed to revert to secondary forest. The value for primary forest is approximate. It was predicted based on allometric relationships derived from trees that were mostly smaller than those in the plot where the primary forest community was surveyed.

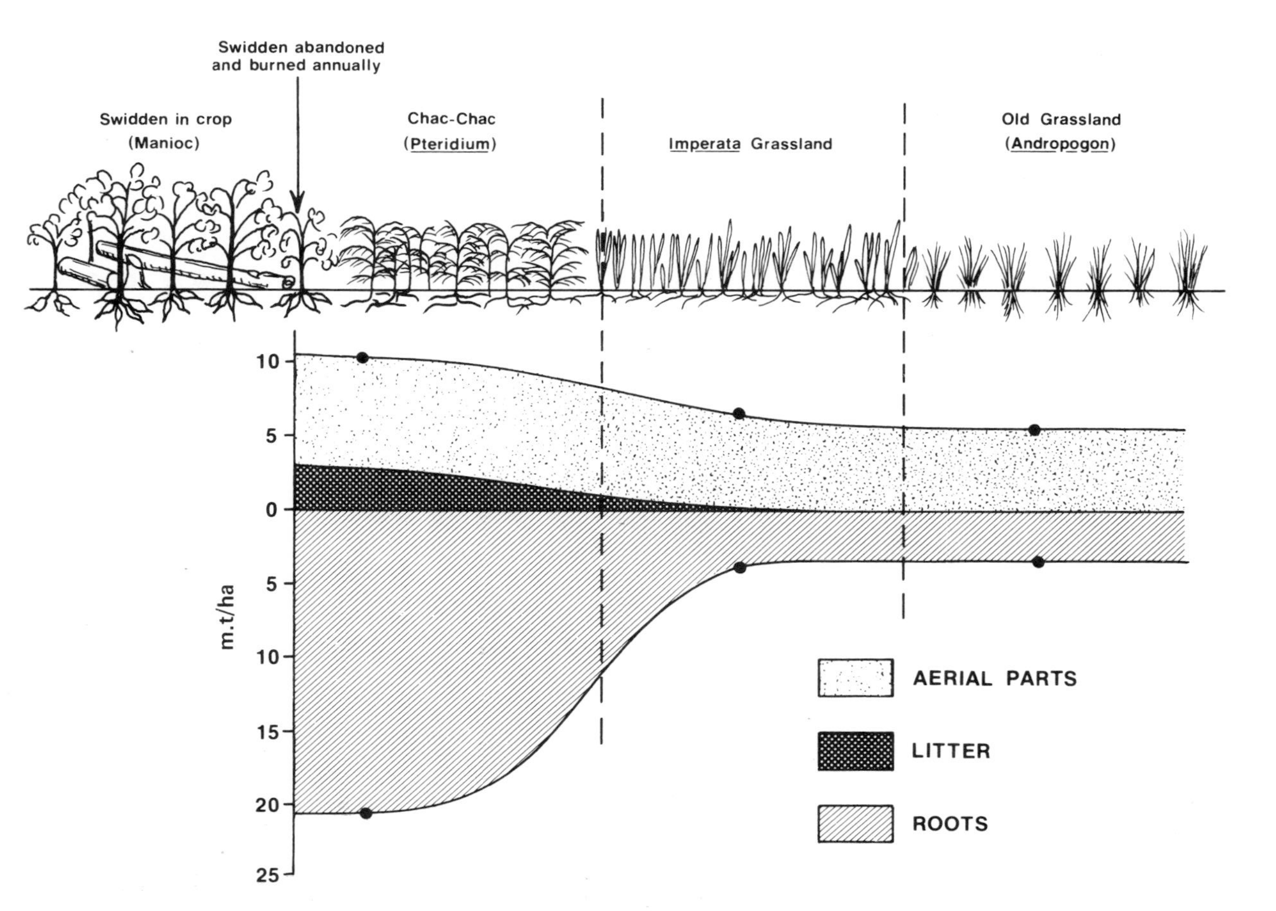

Figure 4.3. Changes in biomass stocks as abandoned swidden changes to old grassland following frequent burning.

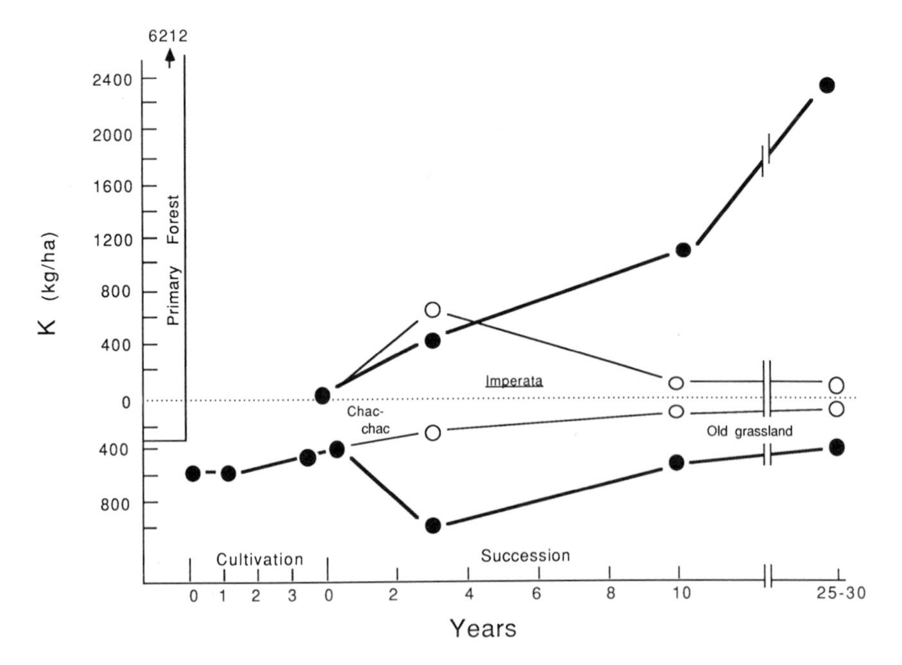

Figure 4.4. Potassium stocks in the vegetation (above 0, *y* axis) and soil (below 0, *y* axis) in three communities: (1) the primary forest (bar at left of graph); (2) cultivated *chacra* or swidden, followed by fire-free secondary succession (solid dots); (3) fire-dominated succession (open circles) with a fern community (*chac-chac*), *Imperata* grassland, and old grassland.

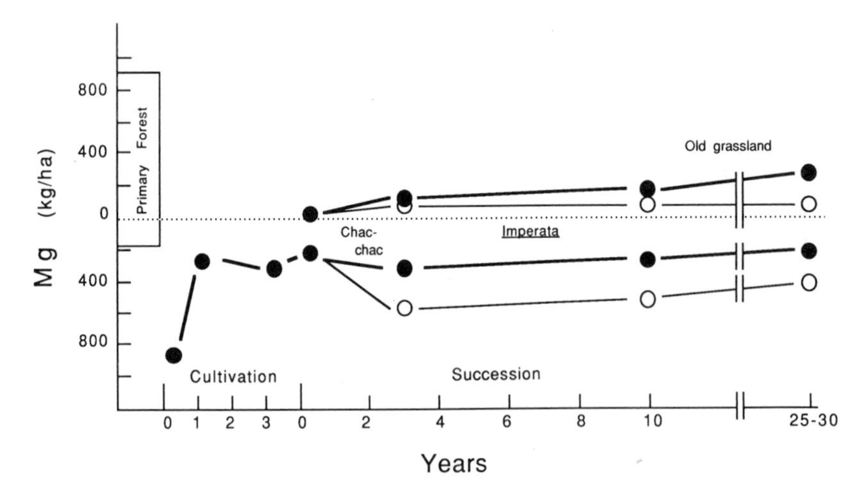

Figure 4.5. Magnesium stocks shown as for potassium in Figure 4.4.

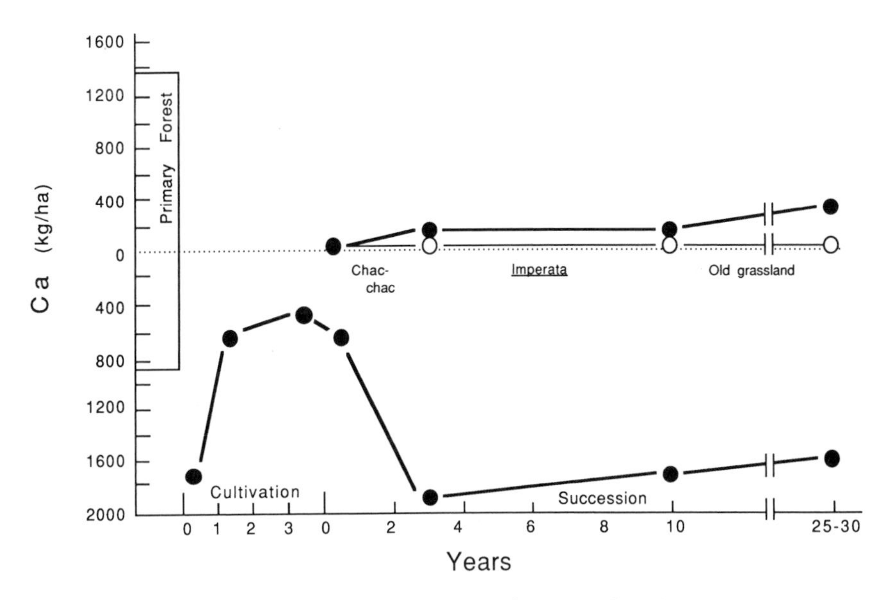

Figure 4.6. Calcium stocks shown as for potassium in Figure 4.4.

Changes in nutrient stocks of the soil and living vegetation during cultivation and during fire-free succession in the Gran Pajonal are shown by the solid dots in Figures 4.4–4.8. During cultivation, stocks of nutrients in vegetation are negligible and there is a trend of decreasing soil stocks for all nutrients. Following abandonment, soil stocks of some nutrients increased for a few years, but over the course of several decades there usually has been a trend of decrease that has paralleled the increase in stocks of nutrients in the successional vegetation. It appears that rebuilding of the nutrient stocks in the biomass during succession occurs partially at the expense of nutrient stocks in the soil. The exception to the trend is available phosphorus in the soil, probably because

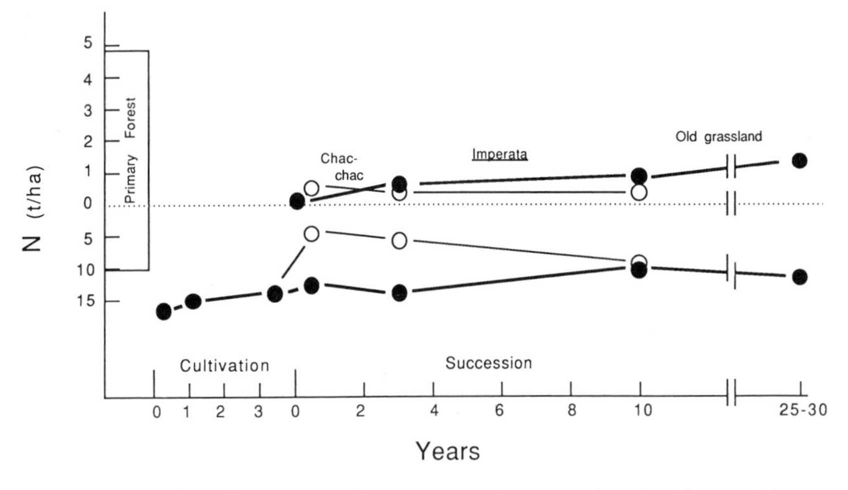

Figure 4.7. Nitrogen stocks shown as for potassium in Figure 4.4.

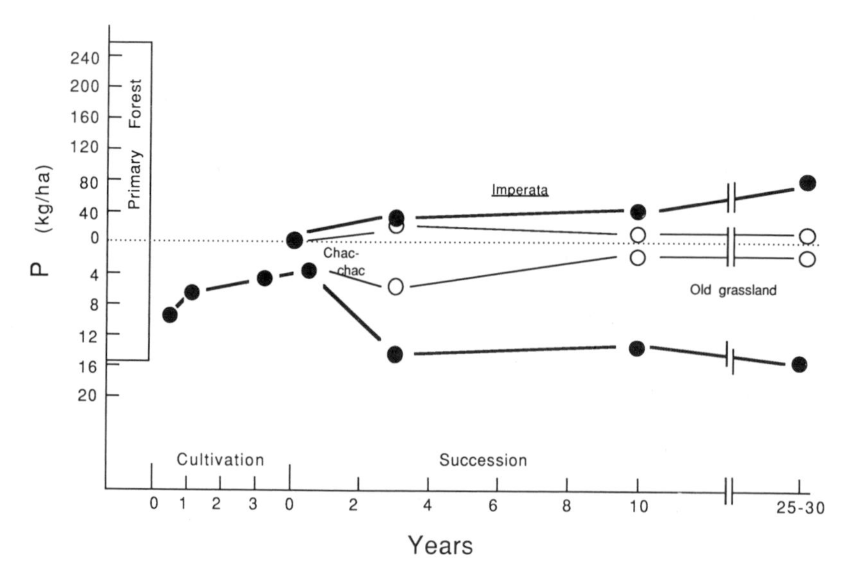

Figure 4.8. Phosphorus stocks shown as for potassium in Figure 4.4. Phosphorus in soil is "available."

available P is only a small proportion of total P and because of pH-mediated shifts in the pools of available phosphorus.

There is no clear-cut rule regarding the necessary length of fallow before cultivation is again possible. Indians rarely cut *chacras* from secondary growth of under 10 years, and 15 is a more common minimum. However, colonists recently moving into the area insist that 7-year secondary growth is perfectly reusable. One stand cut for agriculture had fallow of only 3 years. It seems that such young stands are cut either through laziness or because they occur in sites very handy in relation to domiciles.

Nutrient stocks in communities that are subjected to frequent burning following abandonment of cultivation are shown by the open circles in Figures 4.4–4.8. Stocks of nitrogen, phosphorus, and potassium are all lower in the burned sequence in both the vegetation and the soil. Calcium is lower in the vegetation, but no data are available for calcium in the soil. Only magnesium averages higher in the soil, but the differences may not be significant.

Resiliance of Forest

Recovery of forests from slash and burn agriculture and eventual establishment of primary forests at San Carlos, Venezuela occurred despite very low levels of nutrients in the soil (Chapter 2 and 3). Because stocks of biomass and nutrients in unburned successional forests in the Gran Pajonal are considerably higher than at San Carlos, secondary succession in the Gran Pajonal should progress toward primary forest, uninhibited by problems of soil fertility.

The ability of old grassland to recover to primary forest may be inhibited by relatively low nutrient stocks in the soil. However, some evidence suggests that recovery to forest may be possible. Air photographs taken in 1958 clearly indicate that at that time certain areas of the Gran Pajonal were under grass, whereas in 1976 they were covered with secondary forest. This may reflect reduced burning in areas that have experienced a recent population decrease because of deaths and Indian migration away from the colonist sphere of influence. In any event, it appears that grassland soils in the Gran Pajonal are not so impoverished that a return to forest cannot occur given a fire-free period of many years.

Although eventual growth of forest in burned areas may be possible, recovery of nutrient stocks from amounts in the old grassland to levels of primary forest may take a long time. Potassium is one of the most critical nutrients. The "old grassland" has a total of about 150 kg/ha, compared to about 6500 kg/ha in the primary forest. The input of potassium in rainfall was estimated to be 5.23 kg/ha/yr. If 100% of this input were taken up by the vegetation and there were no input from soil weathering, it would take (6500 − 150 /5.23) = 1214 years for the full stock of potassium to be regained. Because there probably is some potassium input into the soils from weathering of clay minerals it undoubtedly would not take this long, but recovery time almost certainly would be on the order of centuries.

Evaluation of Sustainability

In many parts of the tropics, population pressures are high. Conklin (1959) reports that the land of the Hanunoo in Luzon can support 48 people per square kilometer. Freeman (1955) reports 20–25 in the case of Iban of Sarawak, whereas van Beukering (1947) places 50 per square kilometer as the top level of swidden carrying capacity. Denevan (1971) reports Campa populations of one per square kilometer for the Gran Pajonal, whereas during this study in the more densely inhabited heartland of the area population was between two and three per square kilometer. De La Marca (1773) reports population figures for certain areas of the Gran Pajonal that clearly suggest the 18th century population was greater. Despite these figures, it is unlikely that the more densely populated areas had more than 10 people per square kilometer. In any event, the maximum carrying capacity of the Gran Pajonal has probably never been reached and, particularly today, population pressures do not necessitate the clearing of very young secondary growth with resultant environmental degradation.

Shifting agriculture by the Campa appears to be sustainable without commercial fertilizers when the fallow is 15 years or more and no fires occur. A 15-year fallow does not build nutrient stocks back up to the level of the primary forest. However, it does appear to make available most of the nutrients necessary to sustain a 2–3 year cycle of agriculture.

The current shortening of the fallow period because of the influx of colonists to the Gran Pajonal probably marks the beginning of a trend toward agriculture that is not sustainable, at least without fertilizer supplements. The very short fallows considered sufficient by the colonists reflect the limited time span over which they have cultivated their plots. The nutrient dynamics illustrated in this case study strongly suggest that only a very few such short fallows will seriously deplete the soil nutrients. Agriculture without chemical fertilizer will then not be possible for many decades.

5. Deforestation for Sovereignty Over Remote Frontiers*

Case Study No. 4: Government-Sponsored Pastures in Venezuela Near the Brazilian Border

Robert J. Buschbacher

National boundaries in the Amazon Basin region are sometimes rivers, sometimes watershed divides, but often there are no such natural delimitations. The frontier may be only an unmarked line running through the middle of the rain forest.

Border disputes between adjacent countries are not uncommon in the Amazon region. River boundaries shift as new channels are cut and old ones are filled. Watershed divides change as new streams are discovered. Unmarked boundaries through rain forest are vague, and claims are easily challenged. Consequently, illegal immigration and settlement are easy and often are of concern to the government of the country illegally being settled. For this reason, governments often are anxious to establish a national presence near a frontier. Establishing a presence is an act of sovereignty and can serve as a precedent in any border disputes.

Requirements for frontier establishment in the Amazon are not great. A national development activity can involve cutting the rain forest on a scale only slightly greater than that of shifting cultivation, and the beginning of some type of activity clearly identifiable as a national activity. Pasture development on the

---

* Except where otherwise cited, descriptions, data, and findings in this chapter are from: Buschbacher, R. J., 1984. *Changes in Productivity and Nutrient Cycling following Conversion of Amazon Rainforest to Pasture*. Ph.D. dissertation, Institute of Ecology, University of Georgia, Athens, Georgia, USA.

scale of 10 ha or more can be an ideal choice. It requires an investment that is low on the scale of government spending. Once the forest is cut and burned, and the pasture grass planted, little further investment needs to be made. The actual production of cattle may be irrelevant. The case study described in this chapter is an example of development for frontier establishment near San Carlos de Rio Negro, in the Amazon Territory of Venezuela.

Background

Circumstantial evidence points to the following antecedents for the development project. In the 1970s, due to oil exports, Venezuela was a relatively wealthy country compared to Colombia and Brazil. It was difficult to entice Venezuelan citizens to live in remote border areas such as near San Carlos. In the Amazon Territory of Venezuela, the standard of living was relatively low. Venezuelans living in the Amazon Territory often emigrated to find steady employment and a higher standard of living in cities to the north. As Venezuelans left the region around San Carlos, Colombians and Brazilians migrated across the borders and established their own sites of shifting cultivation within Venezuela. The Venezuelan government may have been concerned about the impact of this settlement. There was already a prolonged disagreement with Colombia about ownership of several islands upstream from San Carlos. Establishment of a "development project" that was clearly Venezuelan apparently seemed to be desirable.

In 1975, an agricultural development credit agency of the Venezuelan government (Fundo Desarrollo) initiated a pasture project near San Carlos. San Carlos is 70 km north of the Brazilian border (Fig. 2.1), not really close to that frontier. Because records of the development agency were not available to us, we can only speculate as to why San Carlos was chosen. One reason may be that many of the Brazilian immigrants were settling in and around San Carlos. Further, San Carlos is the village closest to Brazil where there is a clearly identifiable Venezuelan political infrastructure.

A consortium of 10 San Carlos families was formed and given a 20-year, low-interest loan of about $150,000 US. The original plan worked out between the San Carlos consortium and the development fund called for clearing the forest by tractor; applying fertilizer; seeding *Panicum maximum*, a high-yielding grass commonly used in the Amazon for forage; and construction of three metal water tanks.

As frequently occurs in remote regions of the Amazon, severe logistical problems arose and it was not possible to carry out the project as originally planned. Clearing was done by hatchet and machete instead of by tractor. Fertilizer was not used. Seeds of *Panicum maximum* were replaced with locally obtained tillers of *Brachiaria decumbens*. Water was not pumped but was naturally flowing river water. Between October and December 1979, pasture development was initiated by cutting a 40-ha area of rain forest, some of which was mature primary forest and some of which was 15-year-old successional forest established after shifting cultivation. The consortium contracted out the clearing work to local laborers at the rate of $117 US per hectare for mature forest and $93 US per hectare for secondary forest. After the cut forest was allowed to dry for 2–3 months, it was burned in February 1980. Local observers assessed the burn as intensive and effective. In March and April of 1980, the burned area was planted with clumps of *Brachiaria decumbens* that were collected locally.

In the fall of 1980, the president of the San Carlos pasture consortium spent 3 months in an effort to obtain cows for the pasture. He negotiated a price of $38,000 for 60 pregnant females plus two bulls. Forty percent of the cost was for transport of the cattle by river to San Carlos. Before the cattle were shipped, however, the deal was negated because of demands for political endorsement that were unacceptable to the president. The pasture was then abandoned by the consortium.

At the time that this pasture establishment was occurring, there was a field station at San Carlos that served as headquarters for an ecological research project on the structure and function on the rain forest in the region. One of the objectives of the project was to determine the effect of slash and burn agriculture on productivity and nutrient cycling, as described in case study No. 1 in this book. Because of the presence of a scientific infrastructure at San Carlos, it was decided to take advantage of the pasture establishment to conduct a study of the effect that this type of development would have on nutrient cycling, productivity, and resilience of the rain forest. Permission was obtained from the consortium to carry out the study on their pasture, and several cows were borrowed from local villagers and put on the pasture so that the effect of grazing could be determined.

Methods

Site Description

The pasture was located about 1 km south of the village of San Carlos de Rio Negro, and just a few hundred meters east of the river. The terrain at the site is flat to slightly rolling. The undisturbed forest community surrounding the pasture sites is dominated by two leguminous tree species, *Eperua purpurea* and *Monopterix* sp. The forest is noticeably larger than forests occurring on the Oxisols and Spodosols of the region. The canopy extends up to 35–40 m, and there are numerous trees close to a meter in diameter. A conspicuous mat of litter and partially decomposed humus mixed in a network of fine roots covers the soil surface.

The soil type at the site is an Ultisol (Dubroeucq and Sanchez 1981). The surface soil is very dark brown and consists of 90% fine sand. Between 20 and 135 cm depth, the soil color gradually lightens and the clay content increases. At 3m depth, clay increases to 26%. In comparison with other soil of the Amazon Basin the soil in undisturbed forest is higher in aluminum, about average in organic matter and extractable phosphorus, and below average in

Table 5.1. Comparison of Soil Fertility Parameters in Mature, Undisturbed Rain Forest on Ulitsol in the Amazon Territory of Venezuela to the Range Found in Amazon Soils

Parameter	All Amazon Soils[a] Range	Extent (%)	San Carlos Ultisol
Soil pH	<5.3	81	3.98
	5.3–7.3	19	
Al saturation (%)	0–10	17	
	10–40	8	
	40–70	16	
	>70	59	96
Organic matter (%)	<1.5	9	
	1.5–4.5	74	2.7
	>4.5	17	
Extractable P (ppm)	<3	57	
	3–7	33	6.6
	>7	10	
Exchangeable Ca (ppm)	<80	46	1.2
	80–800	33	
	>800	21	
Exchangeable Mg (ppm)	<24	38	2.3
	24–97	38	
	>97	23	
Exchangeable K (ppm)	<59	62	5.5
	59–117	24	
	>117	15	

[a] From Cochrane and Sanchez (1982).

exchangeable calcium, potassium, and magnesium (Table 5.1). However, it is higher in nutrient content than the oxisols at San Carlos (case study No. 1).

Undisturbed Control Forest

An undisturbed 0.5-ha area of forest close to the pastures was used as a control for the pasture studies. Net primary productivity and nutrient balance were measured in the control forest at the same time that the pasture studies were done, to isolate the effects of pasture conversion from extrinsic variables, such as weather.

Biomass was estimated in the control plot by determining the number of trees in each size class and predicting the dry weight biomass of trees in each size class with the use of allometric equations derived from measurements of 48 trees (case study No. 1). Root biomass was determined in five root pits, each with a surface area of 0.25 m^2 and a depth of 30 cm. Total basal area of trees greater than 2 cm diameter was 44.9 m^2/ha. Dry weight of plant biomass was 465 t/ha, and 9.1% of the total was roots. Dry weight of humus on top of the mineral soil surface was almost 50 t/ha. Samples of biomass and soil were taken for nutrient analysis. Nutrient concentrations were multiplied by the biomass to determine total stocks of nutrients in the uncut forest.

Wood productivity was determined by precise measurements of the annual diameter increment of 104 trees, and converting diameter increment into biomass increment with the use of allometric equations. Leaf litterfall was determined with 15 wire mesh litterfall traps.

Precipitation input of nutrients and nutrient leaching were determined with a network of precipitation collectors and soil water collectors (lysimeters). Subsamples of water were taken for nutrient analysis and concentrations were multiplied by the volume of the water flux to determine total rate of nutrient input and output.

Pasture Sites

There were two different types of sites within the 40-ha pasture. In one type, the prepasture vegetation was mature primary forest. In the other type, the prepasture vegetation was 15-year-old secondary successional vegetation that had established itself following abandonment of a slash and burn agricultural site. Both types were studied to determine potential differences in productivity resulting from different original conditions.

Barbed wire fencing was erected around a plot in each of the two pasture sites. In the site converted from primary forest the area of the plot was 0.2 ha, and in the site converted from secondary forest the area was 0.31 ha. The plots were studied from 1980 through 1983. Standing stock of nutrients on each plot was measured annually through quantification of mass and nutrient concentration in soil 0–30 cm deep, living biomass above and below ground, plant litter, and unburned forest woody residue (slash). Above-ground primary productivity was calculated as the summation of weeding, grazing, litterfall, and net biomass accumulation. Input of nutrients to the pasture ecosystem was measured in bulk precipitation. Losses were determined through water budget calculations and water collections from lysimeters installed below the rooting zone.

Grazing was initiated on the pasture plots after the grass had become established, 1.5 years after planting. The cattle were mature Zebu (*Bos indicus*), borrowed locally. They were allowed to graze in each of the plots for a few days each month. The stocking rates in the pasture from primary forest was calculated to average 0.66 animal units per hectare for the first year, and 0.89 per hectare for the second. The stocking rate in the pasture from secondary forest was 0.43 animal units per hectare for both years.

Cattle growth and mortality were studied during 2 years for a herd of 12 animals living under similar conditions on a ranch adjacent to the experimental plots. Animal weight was estimated each year by allometry (Odend'hal et al. 1979). Fecal output was estimated through measurements of dilution of yterrbium, an indigestible marker administered to the cattle (Teeter et al. 1979). With the use of forage quality measurements, forage intake was estimated to be

2.4 times dung production (Campbell and Lasley 1975, Minson 1980). Urine excretion rates were measured using a harness and bag collection apparatus.

Results and Discussion

Nutrients

Conversion of primary forest to pasture resulted in little change in total ecosystem stocks of calcium and potassium, whereas magnesium appeared to increase and nitrogen decreased (Figs. 5.1–5.4). More important than the change in total stocks was the change in the distribution of nutrients between ecosystem components. In the primary forest before clearing, most of the calcium, magnesium, and potassium and almost half the nitrogen were in the living plant biomass. After clearing, there was a sharp decrease of these nutrients in the plant biomass. As the large biomass became fragmented or turned to ash, the nutrients moved downward, as indicated by the increase of nutrients in the soil and in the detritus on the soil surface (Figs. 5.1–5.4).

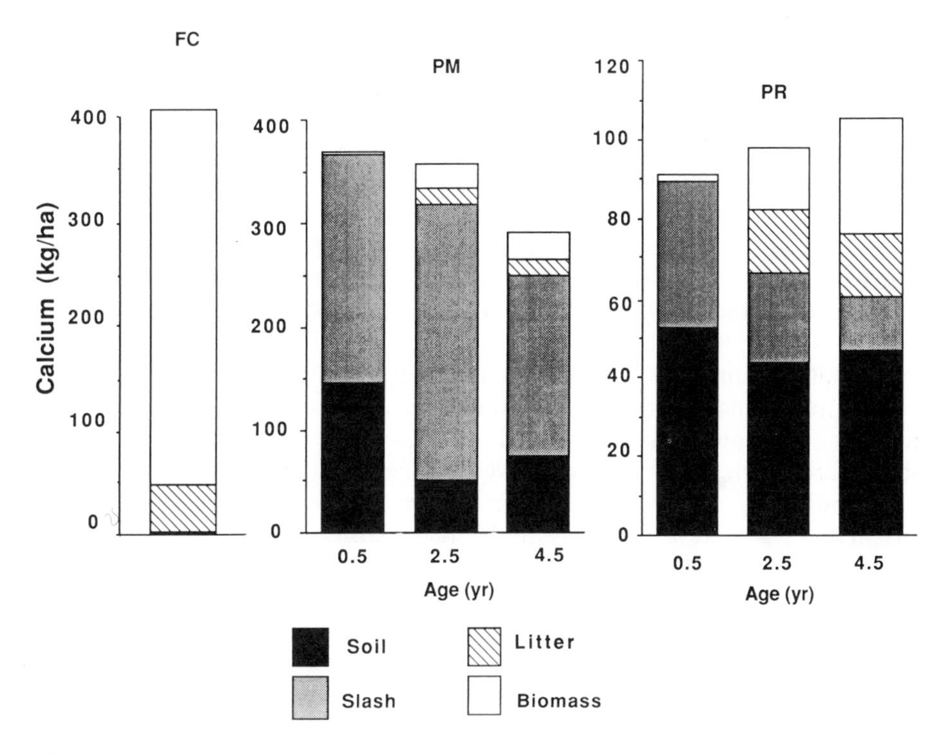

Figure 5.1. Standing stocks of calcium in plant biomass, detritus, and soil of undisturbed, mature rain forest (FC); pasture formed from mature forest (PM); and pasture formed from 15-year regrowth forest (PR) in the Amazon Territory of Venezuela.

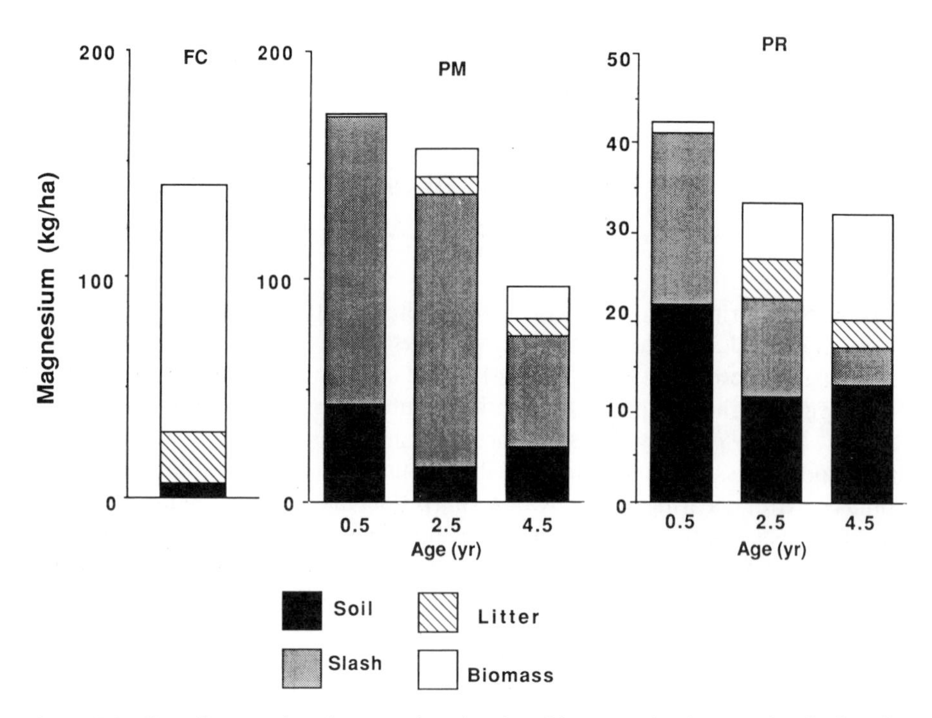

Figure 5.2. Standing stocks of magnesium in plant biomass, detritus, and soil of undisturbed, mature rain forest (FC); pasture formed from mature forest (PM); and pasture formed from 15-year regrowth forest (PR) in the Amazon Territory of Venezuela.

Throughout the 3 years of grazing on the pasture derived from primary forest there was a marked decrease in total ecosystem stocks of calcium, magnesium, and potassium, but no decrease in nitrogen. Stocks of nitrogen, magnesium, and potassium in the pasture formed from secondary forest followed the same trends, but calcium increased. The most important difference between pasture derived from primary forest and that from secondary forest is in the level of soil nutrients at the initiation of grazing. The levels in the pasture from secondary forest were much lower than the levels in the pasture formed from mature forest, indicating a much lower fertilizing effect of the secondary forest.

Generally, a large proportion of the phosphorus in tropical Oxisols and Ultisols is bound with iron and aluminum in an insoluble form (Sanchez 1976) and is not available for uptake by most crop species. In the control forest site, available phosphorus was only 6% of total soil phosphorus. Cutting and burning of the forest raised the amount of soluble phosphorus in the soil (Fig. 5.5). Part of that increase probably came from phosphorus bound in the slash, and part from solubilization of phosphorus bound in the soil. After a year in pasture, levels in the soil quickly decreased (Fig. 5.5). Decline of soluble phosphorus in the soil is probably the most important factor resulting in declining pasture production in the Amazon (Serrão et al. 1979).

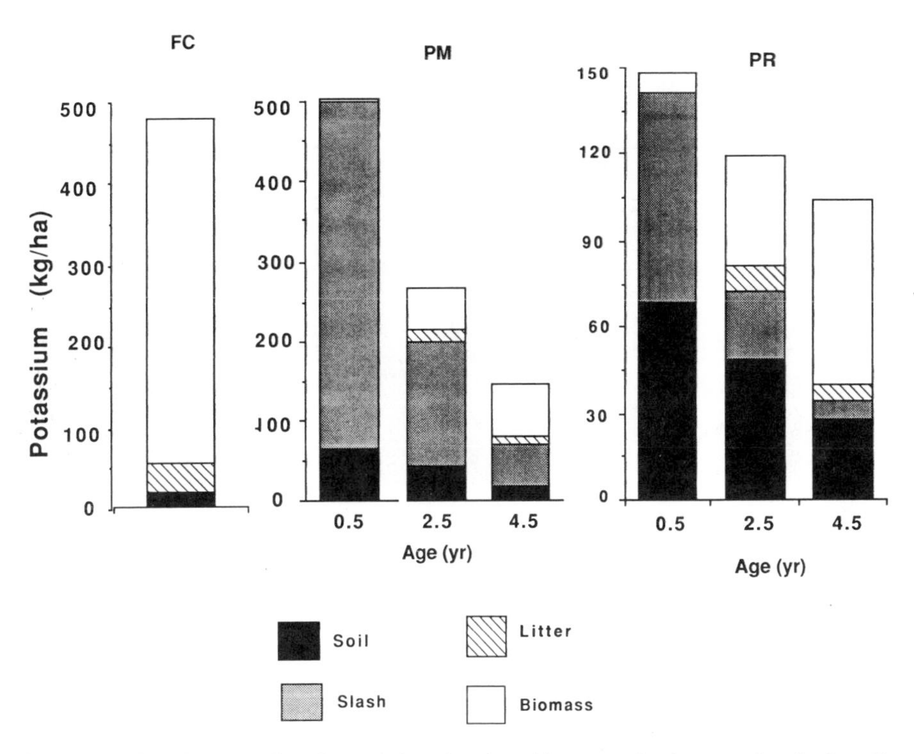

Figure 5.3. Standing stocks of potassium in plant biomass, detritus, and soil of undisturbed, mature rain forest (FC); pasture formed from mature forest (PM); and pasture formed from 15-year regrowth forest (PR) in the Amazon Territory of Venezuela.

Primary Productivity

Total above-ground net primary productivity of the pasture formed from primary forest is very close to the total above-ground, net primary productivity of the undisturbed mature rain forest (Fig. 5.6) during the period of study. The similarity of total production rates suggests a relatively efficient use of the nutrients by the pasture grass. Perhaps this efficiency is a result of the establishment of a fairly dense network of fibrous roots. At 1.5 years, the pasture had 2.05 t/ha of roots less than 2 mm diameter, and 2.48 t/ha of roots larger than 2 mm.

Productivity of *Brachiaria* and total primary productivity in the pasture site converted from mature forest decreased slightly the second year and increased slightly the third. Because the changes are not statistically significant, we can say that the nutrient loss that occurred through the third year in this pasture apparently was not great enough to inhibit productivity.

Above-ground net primary productivity in the pasture derived from secondary forest was about 50–60% of that derived from primary forest (Fig. 5.6).

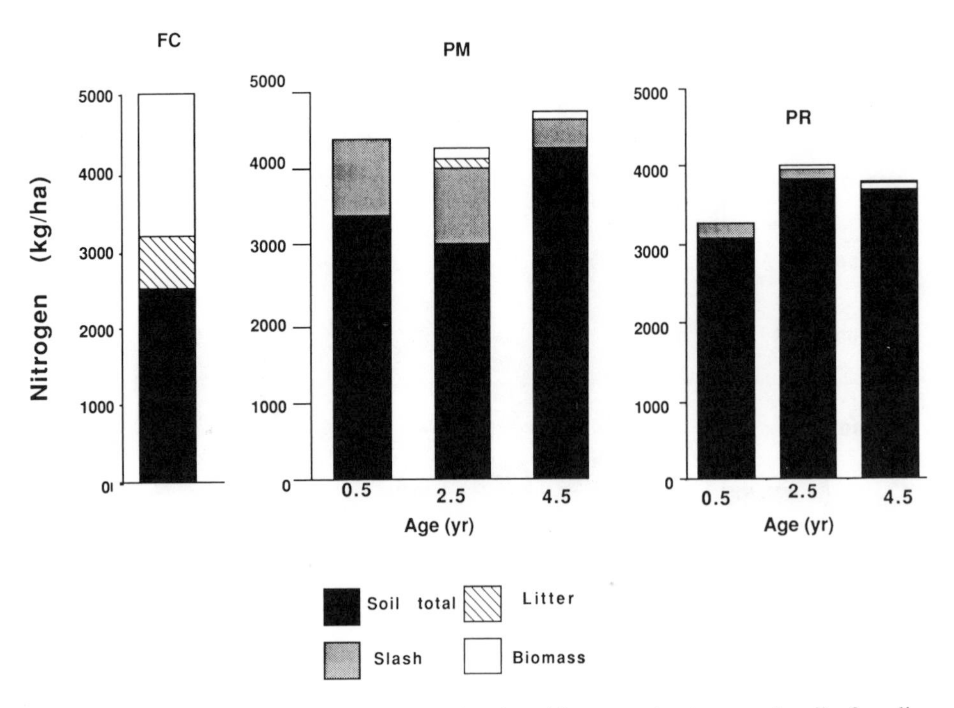

Figure 5.4. Standing stocks of nitrogen in plant biomass, detritus, and soil of undisturbed, mature rain forest (FC); pasture formed from mature forest (PM); and pasture formed from 15-year regrowth forest (PR) in the Amazon Territory of Venezuela.

Levels of calcium, potassium, and magnesium in the secondary forest pasture were only about 25–35% of the levels in the primary forest pasture, suggesting that the low nutrient levels in this site were one possible cause of the low productivity.

Cattle: Their Productivity and Influence on Nutrient Cycles

Cattle detritus as a nutrient cycling pathway differs from litterfall in two ways: the nutrients are more chemically mobile, and they are patchily distributed. Dung and urine produced by cattle make nutrients readily available for plants, thus enhancing productivity. Compared to an ungrazed ecosystem, grazed grasslands have a more rapid turnover of vegetation and nutrients. Grazing accelerates the recycling of nutrients and consequently stimulates primary production. Consumption and excretion by grazers represent important pathways of rapid nutrient cycling (McNaughton 1976).

Although intensive grazing by wildebeests converted senescent grassland to a highly productive site in Africa (McNaughton 1976), there was no evidence of high herbivore productivity at San Carlos. The average growth of the 12 cattle studied in the pasture adjacent to the intensive study plots averaged 9% of the initial weight each year. Average weight gain was 36 kg per individual per year.

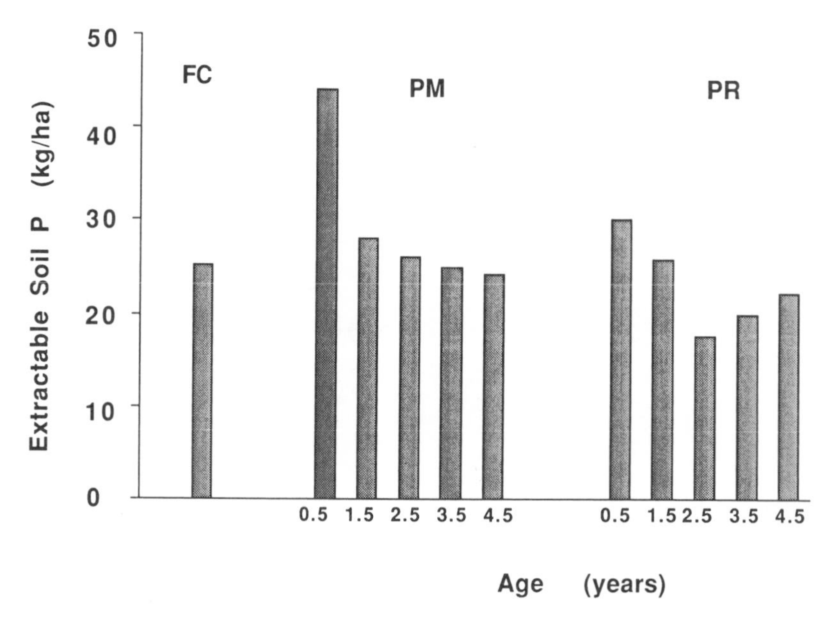

Figure 5.5. Stocks of "available" (extractable) phosphorus in the soil of mature rain forest (FC), pasture formed from mature forest (PM), and pasture formed from 15-year regrowth forest (PR) in the Amazon Territory of Venezuela.

Productivity per hectare was much lower than the average annual liveweight gains commonly reported in the agronomic literature for tropical pastures (Vicente-Chandler et al. 1964, Vicente-Chandler 1974, Stewart 1970). Yields of 149 kg/ha/yr live weight were reported by Toledo and Morales (1979), and 900–1120 kg/ha/yr by Teitzel et al. (1971). An important difference between these latter studies and the one reported here is that the cattle in these studies grazed on chemically fertilized pastures and were fed animal-food supplements. In a study of cattle productivity along the Trans-Amazon Highway, where pastures were not fertilized, productivity of cattle averaged 44 kg/ha/yr (Smith 1978).

At San Carlos, cattle mortality was high. Two animals died from parasites, and three suffered broken bones after falling and were slaughtered. Although the animals are amazingly agile in their ability to climb over and around logs and rough terrain, they apparently have very brittle bones, possibly because of a very low forage phosphorus concentration (0.1%).

In addition to the high mortality rates, there was a complete failure to reproduce during the study period. At least two animals gave birth, but both calves died within weeks, probably from inadequate milk production by the mother. Milk production was so poor that it was not even possible to obtain a sample for analysis. Low milk production probably is caused by low phosphorus levels in the soil and forage.

In summary, cattle growth rates were low, reproduction success was zero, and gains were outweighed by losses from mortality. Forage quantity does not

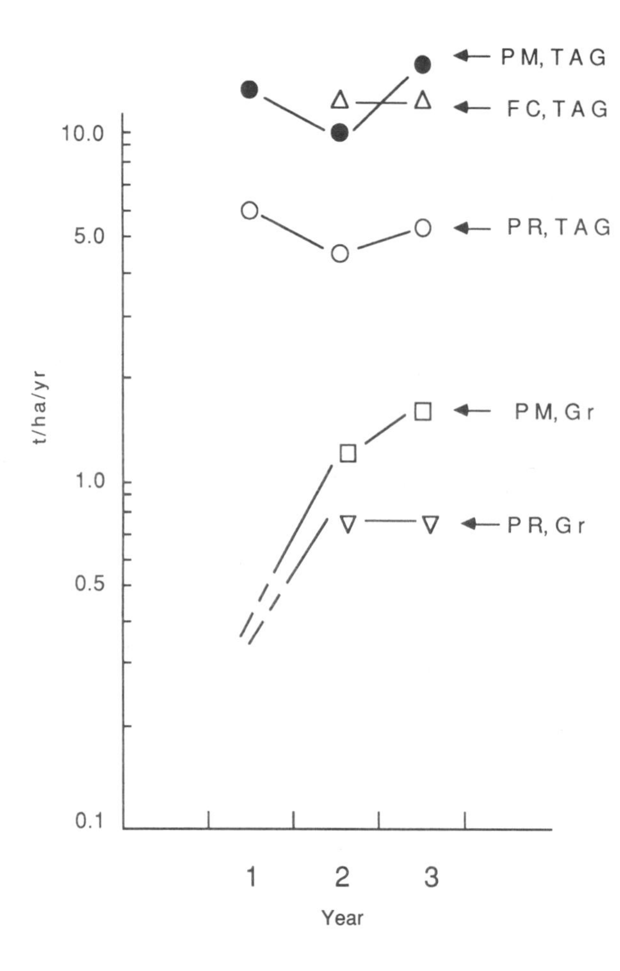

Figure 5.6. Net primary produ
tion per year over a 3-year peri
in pasture formed from matu
forest (PM), pasture formed fro
15-year regrowth (PR), and m
ture rain forest (FC). TAG, to
above-ground production;
production eaten by grazing c
tle. Of the TAG in the pastu
usually about 20% was woo
biomass, whereas in the cont
forest close to 50% was wood

appear limiting and protein content is adequate, but mineral deficiencies probably are critical.

Recovery of Grazed Pasture to Forest

Nutrient levels in the 3.5-year-old pasture from secondary forest are relatively low, and primary productivity in this pasture apparently was affected by the low level. Will the low levels of nutrients inhibit the recovery of this area to forest following abandonment of grazing? This question may be answered by comparing stocks of nutrients at the time of abandonment with the stocks in the slash and burn site (case study No. 1) at the time of abandonment.

Stocks of nutrients in the 3.5-year-old pasture formed from secondary forest were approximately the same or higher than stocks in the shifting cultivation site on Oxisol after 3 years of cultivation (compare Figs. 5.1–5.4 with Fig. 2.3). Lack of nutrients did not appear to prevent succession of woody vegetation, and eventual recovery of mature forest in the abandoned slash and burn site (case studies 1 and 2). Because there were higher nutrient levels in the pasture from secondary forest than in the abandoned slash and burn site, lack of nutrients should not prevent recovery of pasture to forest at San Carlos.

However, lack of seed dispersal into a disturbed area is another factor that can inhibit recovery of forest. Dispersal of seeds into the pasture area probably is slower than into the slash and burn site because of the larger area of the pastures. Nevertheless, the presence of scattered woody seedlings and saplings within the pasture at 3.5 years indicates that at least some tree species can invade and form a cover. As a cover of secondary vegetation forms, invasion by primary forest species is facilitated.

Conclusion

Establishment of lightly grazed pasture, such as described in this chapter, probably is less environmentally damaging than shifting cultivation. Primary productivity and nutrient stocks remained comparable or higher than values for the shifting cultivation site. Following abandonment, it seems likely that the area eventually can revert to primary forest if it remains undisturbed.

From the aspect of cattle production, the project was a disaster. There was a net loss of cattle. Harvests and deaths were not compensated for by reproduction. However, with a greater initial investment it might have become agriculturally sustainable for a time. For example, if salt and antiscour medicine had been given to the cattle, and if the herd had consisted of younger cows, the cattle might have been more productive. However, even with better mineral nutrition there would still be other problems. The climate in the Amazon Basin is not ideal for cattle. Because of the continuous high temperature and humidity, heat and parasite load also may be important. Further, the problems at San Carlos may have been increased because of an overmature herd, the result of logistical and bureaucratic difficulties in obtaining young animals in this region.

From the aspect of frontier establishment, the success of the project is less clear. Certainly the record of the project will be evidence that the Venezuelan government took steps to establish an interest in the region. Whether this will be important in protecting national frontiers in the case of future border disputes remains to be seen.

6. Permanent Plots for Agriculture and Forestry*

CASE STUDIES 5–8

In shifting cultivation as described in the first chapters, fields and crops are rotated or abandoned to fallow for a number of years, and soil fertility is restored through natural processes. Shifting cultivation frequently is considered primitive and predevelopment, despite the evidence that some native systems may be highly sophisticated ecologically and that in such systems species and rotations are well adapted to local conditions and needs (Denevan et al. 1984).

Replacement of shifting cultivation on communal lands with cultivation on fixed plots of privately owned land is frequently considered a first step in development toward integration of a region or a people into the national economy. This chapter focuses on small-scale agriculture and forestry on permanent plots.

The Trans-Amazon Colonization Project: Case Study No. 5

Background

In 1970, a severe drought affected the highly populated northeast region of Brazil. After visiting the region, the President of Brazil, Emilio Garrastazu

---

* Other case study chapters in this book are based on single research projects. In contrast, this chapter has been synthesized from several studies, published in a variety of books and articles. This is because of the requirements of this chapter and the nature of the material available.

Medici, proposed construction of a highway across Amazonia that would enable settlers to migrate and colonize this virtually unpopulated area, thereby partly relieving the agricultural problems of the northeast. Although creating access to mineral and timber reserves and filling an area vulnerable to foreign powers were other reasons for the construction of the Trans-Amazon Highway, the primary motive seems to have been social. In the words of the President, it was "to give men without land a land without men" (Smith 1982, Stone 1985).

To prevent land speculation, the government expropriated a zone 100 km wide along both sides of the pioneer highways. The Instituto Nacional de Colonização e Reforma Agrária (INCRA), an agency of the Ministry of Agriculture, divided a 20-km-wide strip along the Trans-Amazon into 100-ha lots for distribution to incoming colonists. Along the main axis of the Trans-Amazon, a lot extends for 500 m along the road and for 2 km back from the road. Every 5 km, a side road was to be cut through the forest for 10 to 20 km (Fig. 6.1). To encourage small-scale settlers, INCRA offered the 100-ha lots for $700, payable over 20 years with a 4-year grace period. Some settlers were also provided with a four-room wooden house at an additional cost of $100. During the first 3 years of colonization, settlers received six payments averaging $30 a month for a large family to buy provisions and agricultural implements (Smith 1982).

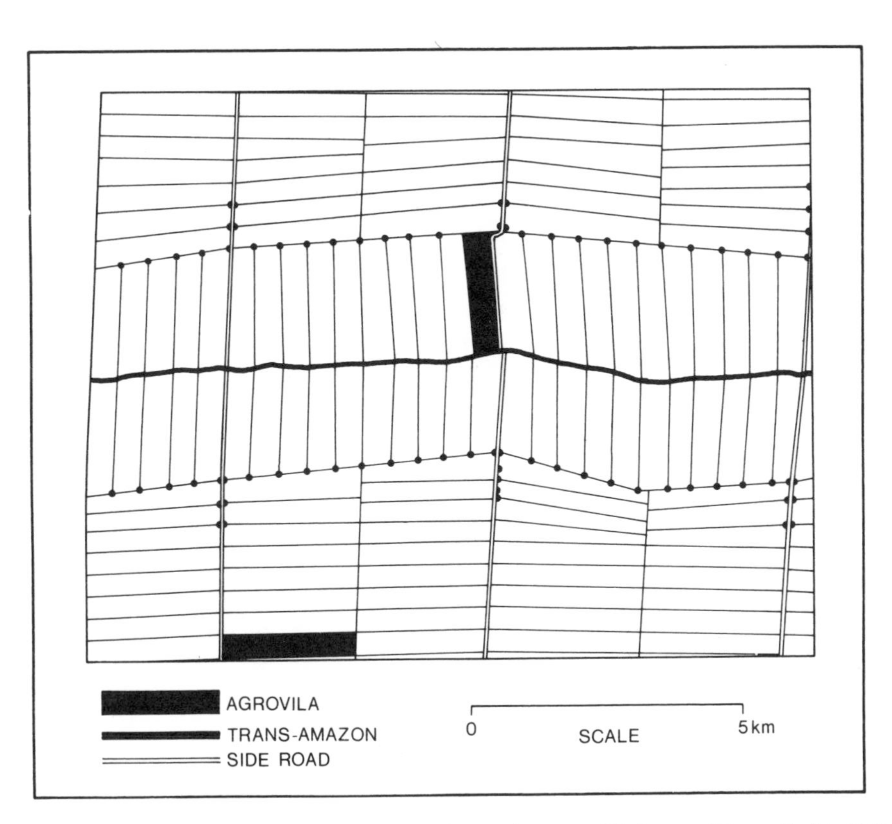

Figure 6.1. The pattern of lots along the Trans-Amazon Highway. (From Smith 1982)

The regional plan was to have a series of agrovilas, INCRA-built communities containing 48 to 66 houses each, located every 10 km along the main axis and side roads (Fig. 6.2). The houses are arranged around a commons, which is used for soccer matches and for grazing livestock. Each community was designed to contain a medical post, a primary school, a general store, and governmental offices. Colonists were encouraged to build their own churches and social centers.

The agropolis, a village for up to 600 families, was the next largest urban center in INCRA's plan. They were to be built every 20 km along the main axis of the highway and were designed as intermediate administrative centers equipped with a small hospital, a dentist, stores, administrative offices, and a police station. The largest city in a region was to be the ruropolis, which was to serve as the administrative center for the population living within a radius of 140 km. It was projected to house up to 20,000 inhabitants and to offer trade schools, banks, hotels, restaurants, a post office, telephones, and an airport (Smith 1982).

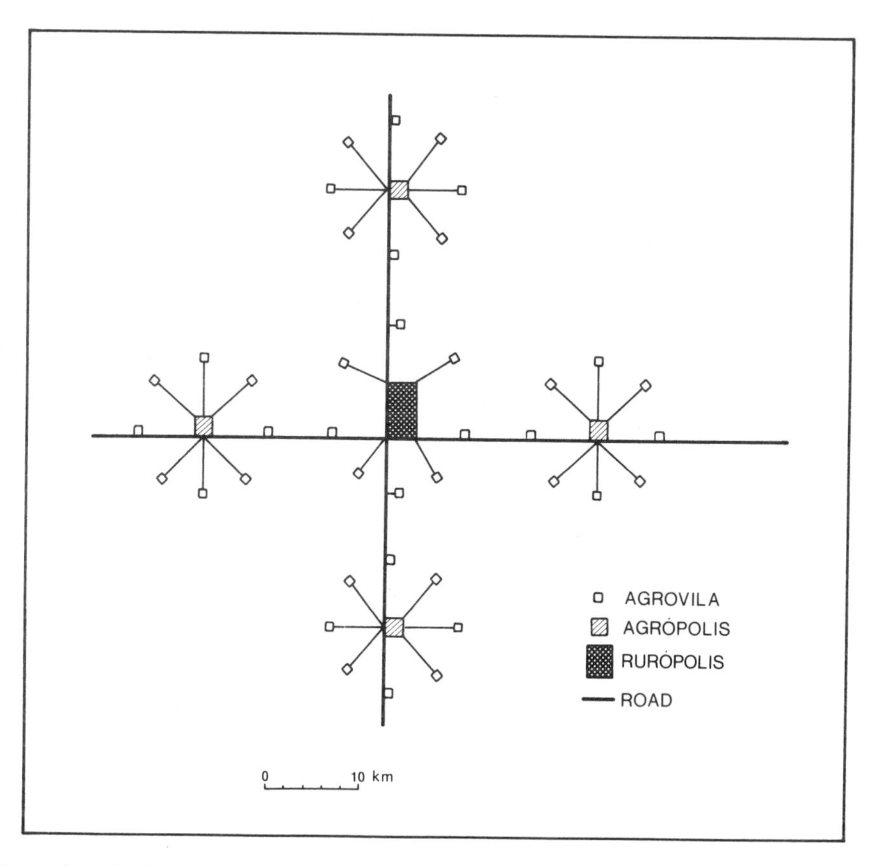

Figure 6.2. INCRA's plan for a rural–urban hierarchy along the Trans-Amazon Highway. (From Smith 1982)

Agriculture Along the Trans-Amazon

One of the centers of colonization along the Trans-Amazon Highway was near Altamira (Fig. 6.3). Much of the soil in the area has been classified as Alfisol (Moran 1981), a soil type considerably more fertile than the Oxisol in the San Carlos study (Chapter 2). For this reason, a larger variety of nutrient-demanding crops was possible than at San Carlos. Nevertheless, the system of land

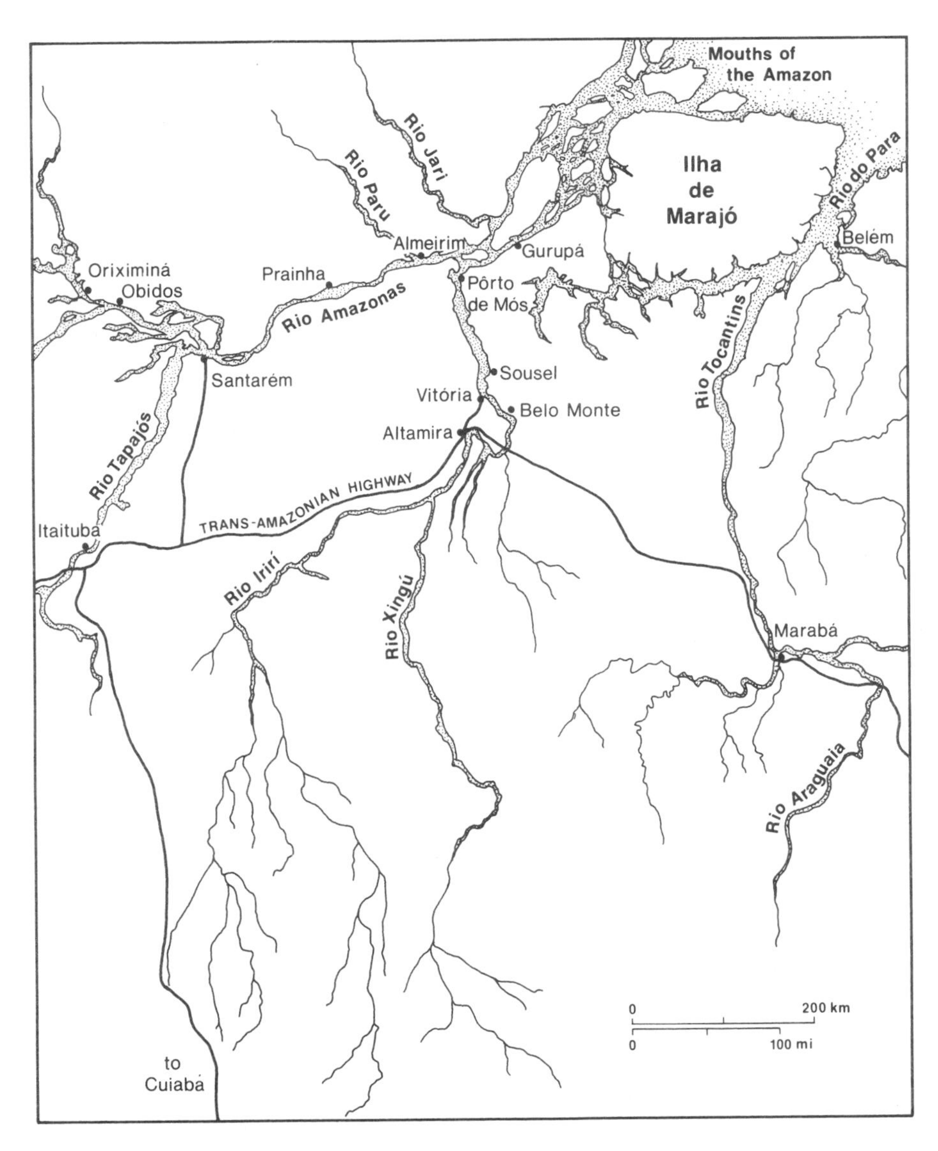

Figure 6.3. Map of the eastern Amazon region, showing Altamira and the Trans-Amazon Highway. (From Moran 1981)

clearing and cultivation near Altimira was still basically slash and burn cultivation.

Moran (1981) has described the land preparation and cultivation techniques. Work was begun early in the drier season of the year, so that the slash could dry and burn before the onset of the wetter months, when planting began. Areas facing the road were cleared first, and each year the new fields moved farther back in the rectangular 100-ha holding. The first step was undercutting, in which small trees, underbrush, and vines were cut. With this vegetation removed, it was easier to cut the larger trees. Most workers followed the traditional method, which is to cut a large tree so that it falls on top of smaller ones that have been partially cut but not toppled. The most common tool was the machete, but a number of colonists owned powersaws and used them in their own work and to cut trees for others.

Timing of the burn was important. The farmers waited until the slash was as dry as possible in order to get the hottest and most complete burn, and consequently the greatest amount of ash deposited on the soil. Productivity, at least the first year, generally was correlated with the completeness of the burn and the quantity of ash. Farmers' decisions as to what to plant were based on past experience, the yields obtained by neighbors, information about crops and prices, evaluation of needs, and in some cases the advice of the extension service. An important decision was whether to plant cereal crops that would mature all at once, or root crops whose production would be available year round. The necessity for plants used for medicinal purposes also was considered.

Seeds of corn and rice and cuttings of manioc usually were planted before the onset of the heaviest rains. Upland rice sometimes was planted on well-drained slopes, whereas corn was planted in low areas, which usually have deeper, less rocky soils. However, sometimes corn was planted in the better drained soils, and rice in the lower areas (M. Dantas, personal communication, 1985). In areas of poor soil, or where the burn was incomplete manioc may have been planted.

Beans, the other major crop, were planted after the end of the rainy season, when the corn stalks began to dry out. Bean seeds were put into the ground at the base of the stalks, which then served as support for the growing beans. Other major crop species were bananas, cacao, and papaya. Some farmers also planted a great variety of edible plants, such as green and hot peppers, collard greens, okra, guava, tomatoes, spices, pineapple, squash, sweet potato, and mango near their houses. Sugarcane sometimes was planted as an ornamental in the front of the house.

Results

Farming success varied significantly. The most successful farmers were generally long-time residents of the Amazon, who were familiar with subtle differences in soil quality, and could detect differences in soil by the type of native vegetation growing on the soil (Moran 1981). Nevertheless, even among the best farmers, yields generally were low. Low productivity has been attributed to a number of causes. Low rice yields have been caused by inappropriate varieties promoted by the extension service (Smith 1982). The most commonly planted variety was developed in Campinas in southern Brazil. It proved unsuited to the climatic conditions of the Amazon Basin because the long stems fell over and became flattened against the ground, where they were easily eaten by rats and doves. Weed competition also affected rice production.

Another important factor in the Trans-Amazon program was the combination of humid climate and transportation difficulties that resulted in crops rotting before they could be gotten to a market. Losses during storage were high because of inadequate protection and storage facilities. Pests and predators also were important factors. Their control in the field was difficult because pesticides were generally not available or too expensive (Smith 1982).

The most important factor in low productivity, however, probably was low soil fertility. Although nutrient availability in the soil increases following cutting and burning of slash, erosion, leaching, and fixation of phosphorus cause a rapid decline in fertility within a few years (Jordan 1985a). Moran (1981) carried out a study of changes in soil fertility during a cycle of cultivation in fields near Altamira along the Trans-Amazon Highway. Three to five areas were sampled from plots under mature forest; plots in newly burned fields; soils in the first, second, and third year of cultivation; and soils under secondary growth. The results (Figs. 6.4–6.6) clearly showed an increase in nutrients available in the soil following burning and a gradual decrease during cultivation and abandonment. The largest changes were in available phosphorus (Fig. 6.6). Phosphorus is the nutrient element that most frequently is limiting to plant growth in the lowland humid tropics (Jordan 1985a).

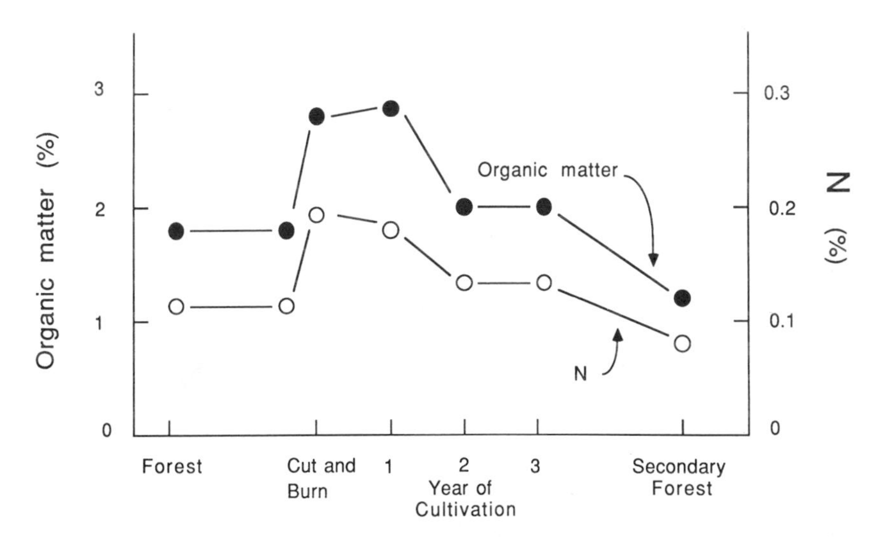

Figure 6.4. Average levels of organic matter (OM) and nitrogen in soils in various stages of cultivation cycle near Altamira. (Drawn from tables in Moran 1981)

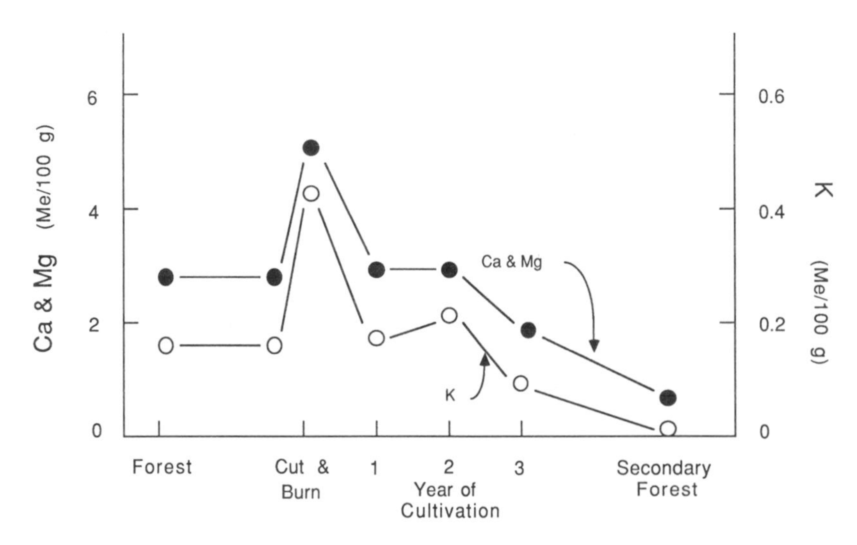

Figure 6.5. Average levels of potassium and of calcium plus magnesium in soils in various stages of cultivation cycle near Altamira. (Drawn from tables in Moran 1981) Me = milliequivalents.

Evaluation

The Trans-Amazon colonization scheme cannot be considered successful (Stone 1985). In all, only 7389 families were established in three model colonies during the 7 years from the beginning of the program until 1977, relative to a goal of 100,000 families by 1974. A major factor in the slowdown of the colonization rate has been the disappointing agricultural yields. Rice yields averaged only 1593 (s = 679) kg/ha. Corn yields averaged only 1320 (s = 646) kg/ha). A

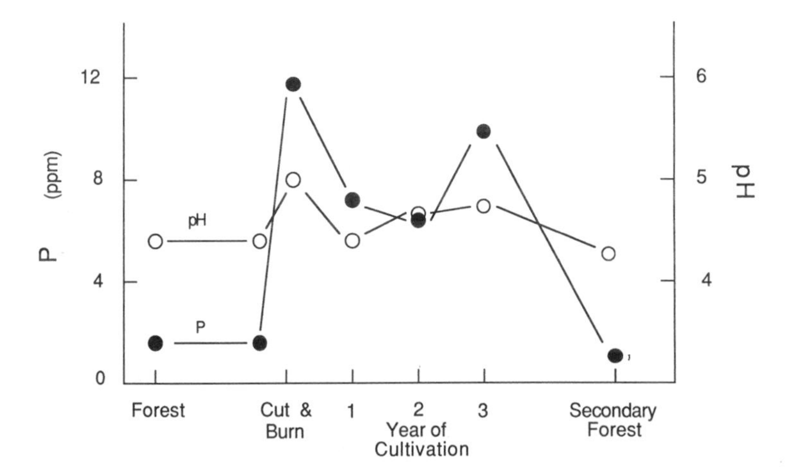

Figure 6.6. Average levels of pH and extractable phosphorus in soils in various stages of cultivation cycle near Altamira. (Drawn from tables in Moran 1981)

typical farmer plants 8 ha of rice every year but only grosses about $1900 from the sale of the crop, a meager income considering the costs of manufactured goods in Amazonia and the wages that must be paid for labor (Smith 1982).

The Trans-Amazon agricultural model was based on techniques developed in and adapted for temperate-zone agriculture. Technological inputs, such as fertilizers and pesticides, were required for adequate production. Transportation and storage facilities were required for marketing the products. Because of environmental conditions very different from the temperate zone, and because the political and economic infrastructure that was to supply the technology to make the agriculture sustainable was not sufficiently developed, the agricultural development program did not achieve many of its objectives (Smith 1982).

Continuous Cropping with Fertilizers: Case Study No. 6

Trans-Amazon agricultural development was generally not successful because the technology and subsidies necessary for sustainability of crop production were not adequately available to the colonists. If necessary technology and subsidies could be made sufficiently available, would it be technologically feasible to sustain continuous cultivation of crops on acid Amazonian soils?

To answer this question, a research program was initiated in 1971 near Yurimaguas, Peru, the westernmost large fluvial port on the Amazon headwaters (Sanchez et al. 1982). The 2100 mm of rainfall at the site is well distributed throughout the year, with 3 months averaging 100 mm, and the rest about 200 mm. The native vegetation is tropical rain forest. The principle soil at the experiment station is a flat, well-drained ultisol, with a sandy loam surface over a clay loam subsoil. It is very acid (pH 4.0), high in aluminum, and deficient in phosphorus, potassium, and most of the other nutrients. These conditions are typical of much of the Amazon Basin (Sanchez et al. 1982).

Sanchez et al. (1983) have described the experiment. Three fields ranging in size from 1 to 2 ha were selected in an area having the same soil, geomorphic position, and standing vegetation. The fields were cleared by the slash and burn method. The felled logs were sawed and removed, but tree stumps were left in place. Each field included a 4 × 7 factorial design with four cropping sequences as the main plots and seven fertility treatments as subplots.

Various fertilization treatments were tried and modified based on soil analyses and crop yield. One of the first treatments was called the "maintenance" treatment and was designed to replace macronutrients taken up by the crops. Until July 1975, 40 kg/ha N as urea, 20 kg/ha P as superphosphate, and 30 kg/ha K as potassium sulfate were used for each annual crop. "Complete" treatments received a base application of 80–100–80 kg N–P–K per hectare and were limed to pH 5.5. Subsequent crops received 100–26–80 kg N–P–K per hectare (Villachica et al. 1976). Along with the fertilized plot there was a control that was planted and weeded but not fertilized.

The complete fertilizer treatments substantially increased the levels of calcium and available phosphorus in the soil and somewhat increased the magnesium and potassium (Fig. 6.7). Treatment also increased soil pH and decreased exchangeable aluminum and aluminum saturation (Fig. 6.8). Aluminum toxicity is a common problem in crop plants in the Amazon basin, and decrease in aluminum availability helps sustain crop growth. Only nitrogen and carbon levels in the soil were relatively unaffected by fertilizer treatments (Fig. 6.9).

The results of the initial trials showed that fertilization improved yield and prolonged the period of good yield. Nevertheless, even the "complete" fertil-

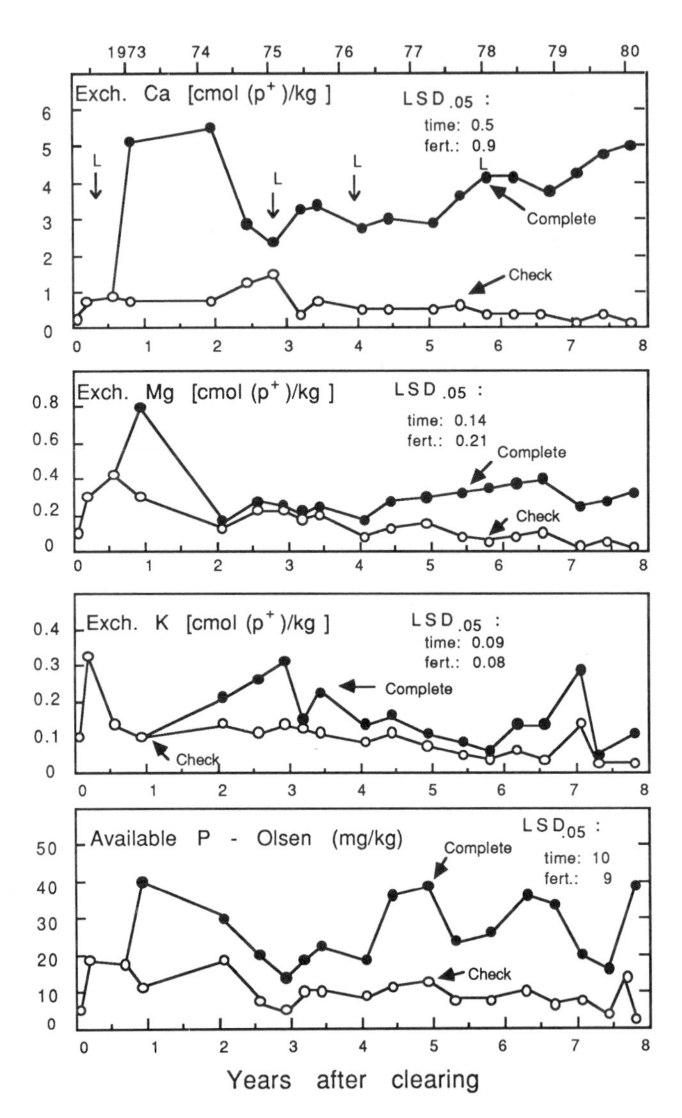

Figure 6.7. Levels of several macronutrients in a plot with "complete" fertilizer treatment and in "check" or control plots. (Adapted from Sanchez et al. 1983) LSD = Least significant difference.

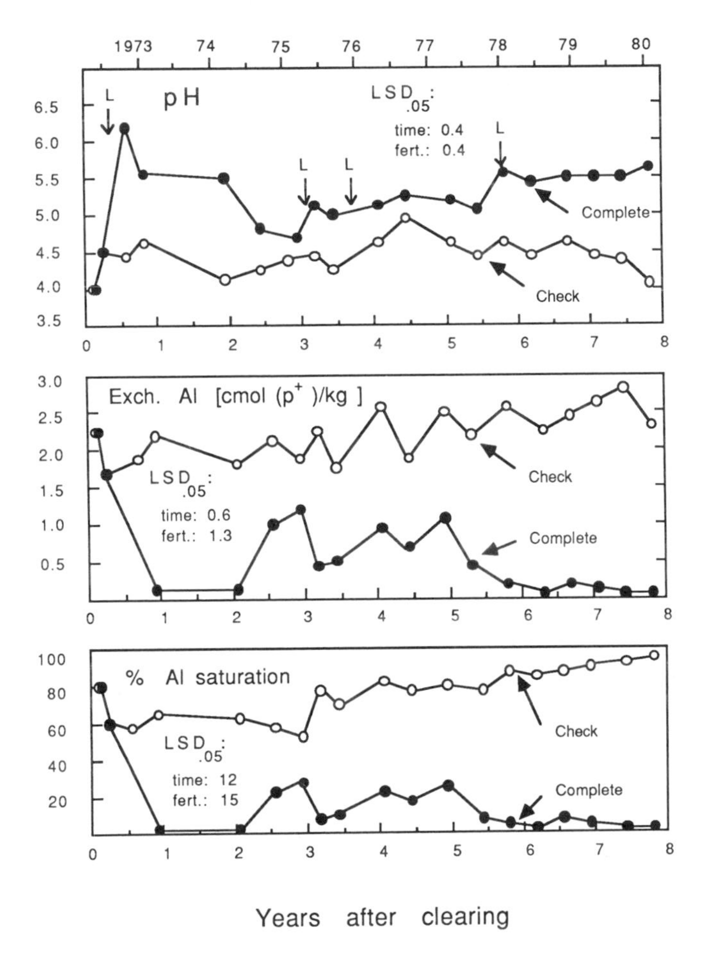

Figure 6.8. Levels of pH and aluminum in a plot with "complete" fertilization treatment and in "check" or control plots. (Adapted from Sanchez et al. 1983). L = time of lime application.

izer treatment failed to halt yield declines over a period of several years (Fig. 6.10).

One of the steps taken to remedy declining production, in addition to increasing applications of major nutrients, was the application of the micronutrients zinc, manganese, iron, copper, boron, and molybdenum (Villachica and Sanchez, 1978, Sanchez et al. 1983). The addition of higher levels of macronutrients and the inclusion of micronutrients reversed the declining trend of yield after 1975 (Fig. 6.11). By 1980, yield had stabilized at levels much higher than those in the unfertilized plots.

Nicholaides et al. (1983) concluded that:

> The understanding of soil fertility dynamics was the key to the development of the successful "Yurimaguas technology." The nutritional needs of the crops in the

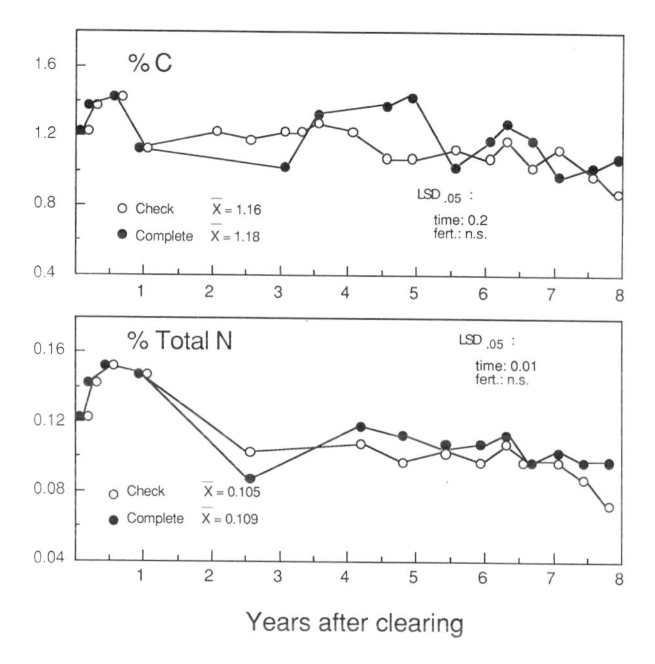

Figure 6.9. Levels of carbon and nitrogen in a plot with "complete" fertilization treatment and in "check" or control plots. (Adapted from Sanchez et al. 1983)

> Amazon Basin, as with crop production anywhere, can be determined only by continual monitoring of soil fertility dynamics through soil and plant sampling and testing. Only then can the most judicious use of lime and fertilizers for crop production be ascertained. [p. 125]

Economic Considerations

The experiment at Yurimaguas demonstrates that if the technology and subsidies are available, sustained agricultural production is possible on acid Amazonian soils. However, the economic feasibility of intensive fertilization may be limited to areas with a well-developed infrastructure. The scientists involved in the project stated:

> The socioeconomic conditions [of the Amazon region] provide another limitation to the widespread adaptation of the Yurimaguas technology. The Yurimaguas area, though not a privileged region of the Amazon, has an unpaved road to Lima and several rivers that link it with markets in the rest of the country and beyond. (Nicholaides et al. 1983, p 139).

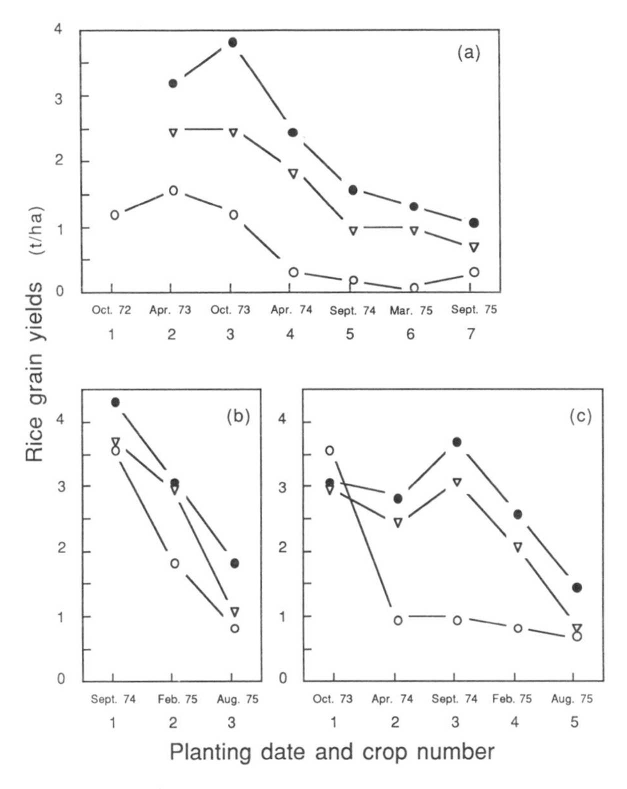

Figure 6.10. Yields of a continuous rice cropping system in eastern Peru as a function of time since clearing: (a) 1972, (b) 1974, (c) 1973. Solid dots, "complete" treatment; triangles, "maintenance" treatment; circles, unfertilized. (Adapted from Villachica et al. 1976)

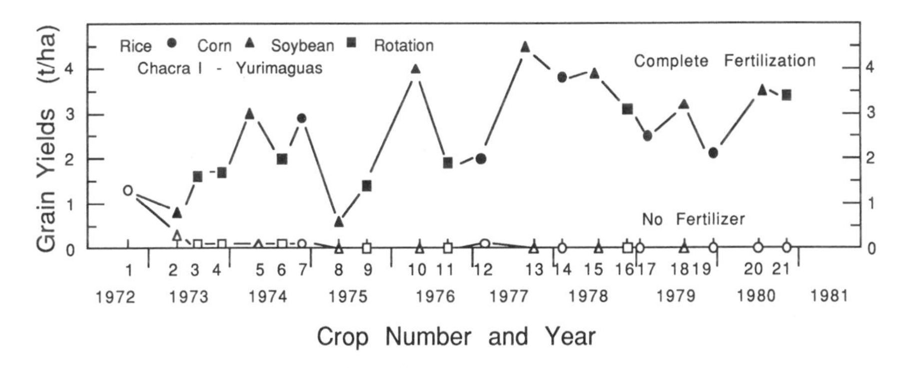

Figure 6.11. Yield record of a continuously cultivated plot on an Ultisol of Yurimaguas, Peru, with and without complete fertilization and lime. (Adapted from North Carolina State University, 1972-1981).

Agroecology at Tome-Assú, Brazil: Case Study No. 7

Heavy application of commercial inorganic fertilizers, as at Yurimaguas, Peru, is not the only method of sustaining agricultural productivity in the Amazon Basin. Agroforestry or agroecology are other possibilities. These terms are applied to agriculture that attempts to minimize requirements for commercial fertilizers and pesticides through innovative rotations and combinations of crops. The description here of the agroecology project near Tome-Assú, Pará, Brazil, is based in part on a visit to the central cooperative and surrounding farms by C. Uhl and C. Jordan in 1984.

Background

In the early 1900s, groups of Japanese peasant immigrants established agricultural colonies at several sites in the Amazon Basin (Sioli 1973). One of the colonies was located on the Rio Acará-pequeno, 150 km south of Belém, and named Tome-Assú. The colony initially tried to raise cacao, but this effort failed (Sioli 1973). The colony began to decline, and the process was accelerated by a severe malaria epidemic that killed many Japanese settlers and caused others either to move to southern Brazil or to go back to Japan. Low fertility of the soil also probably played an important role in the lack of agricultural success.

About 1930, a Japanese immigrant brought seeds of black pepper (*Piper nigrum*) to Tome-Assú, and this crop proved more successful. During World War II, when the Brazilian government transformed this Japanese colony into an internment camp, the settlers started to expand the pepper plantations. By the end of the war, the product, which was new for Amazonia, was available to the Brazilian market and soon also to the world market (de Oliveira 1983).

However, as plantations of black pepper age they become increasingly susceptible to the fungus *Fusarium solani* (Nascimento and Homma 1984). It became necessary for the Japanese to explore other crops to maintain the colony. During the decades after the War, the agriculture became increasingly diversified (Sioli, personal communication, 1985). By 1984, the colony had developed an ecologically complex system of agriculture and had evolved into a series of individual farms cooperating to increase efficiency of production and minimize the necessity of importing fertilizer and other agricultural subsidies. The effort was managed by a cooperative, which also served to coordinate marketing activities (de Oliveira 1983).

Management Principles

The farmers had no formal list of procedures that they followed in order to maximize their reliance on natural processes and minimize reliance on imported subsidies. However, observations of their techniques and field plots led to a compilation of the following management principles, which they seemed to apply whenever possible.

1. Utilize, as much as possible, tree species in which the biomass harvested, such as fruit or latex, is only a small proportion of the total plant biomass, thereby minimizing nutrient loss from the system during harvest and minimizing disturbance to the soil.
2. For annual crops, use only species of high economic value and plant them only once, or twice at the most. Immediately follow with species of lower economic value but that are less demanding of soil nutrients. When possible, use species that can both enrich the soil and provide economic benefit.
3. Maintain complete ground cover as much as possible to minimize erosion and deterioration of soil physical properties.
4. Maintain as high a diversity of crops in an area as possible, to exploit soil nutrients and sunlight fully and to inhibit problems of diseases and insects. Diversity of species also maintains a more stable economic income.
5. Recycle, as much as possible, both animal and vegetal organic matter into the soil.

Examples of Agroecosystems at Tome-Assú

A cycle may start with the clearing of secondary forest growing on a site thought ready for cultivation. Some trees may be used for lumber, and others converted to charcoal. After the plot is cleared, remaining slash is burned, and seedlings of rubber trees tolerant of the low-fertility soils may be planted at intervals of several meters. For several years, before the canopies of the rubber trees close, other crops are planted to take advantage of the freshly fallowed soil and to maintain soil cover. First, corn is planted between the rubber seedlings. When the corn is half a meter tall, ginger is planted. When the corn is harvested, the stalks and leaves are collected and this mulch is spread around the base of fruit tree seedlings in neighboring plantations to conserve moisture and improve soil quality. Following corn harvest, cotton, winged beans, and peanuts are planted in the same plot. These crops do well because of the fertilizing effect of decaying corn roots. When the ginger is harvested, the husks, which the farmers claim repel nematodes, are spread on nearby beds of onions, which are susceptible to the nematodes. If a second rotation of corn–beans–cotton is carried out, some fertilization is required.

After 2 years, the canopy of the rubber trees begins to close. A high-yield variety stem of rubber is grafted to the root stock, and later, a second, fungal-resistant graft may be added. Nonresistant leaves in Amazon rubber plantations usually will be killed by attacks of the leaf fungus *Dothidella ulei* (Sioli 1973).

Pepper is still an important crop at Tome-Assú. It can be cultivated for almost 10 years before problems with rot force abandonment. Pepper is a vine, and sometimes living trees of nitrogen-fixing species are used to support the vines. Palm species, valuable for heart of palm and for fruit used for juice and as pig food in neighboring farms, often are seeded in naturally between the pepper plants by parrots and macaws, and these palms may be left to grow. Several years before a pepper plantation is abandoned, trees species with high commercial value for wood may be planted among the pepper vines. When the pepper plants are abandoned, therefore, the plot is already supporting a vigorously growing forest of valuable species.

Cacao also is an important crop, now that techniques to control disease have been developed. Cacao is an understory tree, and a variety of species, such as valuable timber species, nitrogen-fixing trees, rubber, and coconut palms, are used as an overstory. To control fungal disease of cacao, the fruit is transported out of the plantation before the seed is extracted. Uninfected pods are composted and then used as organic fertilizer in the cacao plantation or in other fields. Infected pods are burned and the ash is used as fertilizer. Commercial fertilizer also is sometimes used. To gauge the amount of fertilizer that will be required during a season, the farmer will inspect the plantation carefully during the flowering season. If a large number of flowers are observed, a large crop is anticipated and heavy applications of fertilizer are made.

Disease in cacao plantations also is controlled by pruning branches every year, which increases wind flow and decreases humidity. Pruning also is said to stimulate fruit production and aid in harvest. In some plantations, vanilla, a vine, is planted at the base of cacao trees, which provide support. Vanilla is an extremely high-value product but pollination, which is done by hand because natural pollinators are lacking, is a very tedious process.

Besides cacao, a variety of other fruit trees is grown at Tome-Assú. A nursery is maintained for starting seedlings of fruit trees before they are planted in the field. Enriched and sterile soil from beneath incinerators is used as a medium for establishing seedlings in the nursery.

Several of the farms belonging to the cooperative at Tome-Assú concentrated on animal production. In one, some 13,000 chickens produced 30 tons of organic fertilizer per year. Husks from rice grown on another farm are spread on the floor of the chicken house. Every three months, the husk–manure is bagged up and used to fertilize plantations of fruit trees. At another farm, pigs were fed corn, rice, and manioc grown on neighboring farms and supplemented with minerals. Organic waste from the pigs was used to fertilize pepper plantations.

These are just a few of the combinations of crops at Tome-Assú. Other crops that we observed there were pineapple, oil palm, taro, several types of citrus, papaya, and plantain.

Economic Considerations

Although some commercial fertilizer was bought and used, a representative of the cooperative claimed that proceeds from the sale of crops more than paid for the cost of the fertilizers. Even it this is true, the Tome-Assú model, like the Yurimaguas model, may not be applicable in areas with less developed transportation. Tome-Assú is accessible by road to Belém (1 day) and Sào Paulo (4 days), which is a big advantage in marketing many of the high-value crops. Such accessibility probably is a decisive factor in the profitability of the operation.

Another factor that may be important in the economic success of Tome-Assú is cultural discipline. The work may be tedious and requires a type of discipline that occurs in traditional Japanese society but is not common in other cultures.

Sustained-Yield Forestry: Case Study No. 8

Despite many variations, forest harvest techniques can be divided into two broad categories, selective cutting and clear-cutting. A clear-cut forest can regrow naturally, or it can be stocked artificially. A clear-cut site that is regrowing, or that is in the early stages of plantation, is in a sense "out of production"; that is, it has been set aside to remain undisturbed until the trees are ready for harvest. A rotation of clear-cut forest plots is similar to shifting cultivation, in that there is a long period of fallow between harvests.

In contrast, selective harvest forestry is similar to permanent plot agriculture, in that the same site is used on a more or less continuous basis. It is frequently called sustained-yield forestry. Trees are harvested periodically but only a portion of those on a site are taken. Trees may be selected on the basis of size, species, form, or other criteria. Remaining trees may be left untouched, they may be killed, or they may be encouraged by removing competing trees.

During the first half of the 20th century, selective harvest management systems for naturally occurring tropical forests of Africa and southeast Asia were developed (Baur 1964). In Africa, the systems were often referred to as the "tropical shelterwood system" (Fox 1976). These systems were designed to promote the establishment, survival, and growth of the seedlings and saplings of desirable species by the poisoning of undesirable trees and the removal of vines and weeds. After several years, when surveys showed that the reproduction of desirable species was well established, the canopy trees of the timber species were harvested. The tract was then monitored, and weedings and thinnings took place as needed. The system was abandoned in the 1960s, partly because it did not make sufficiently intensive use of the land to compete with other forms of land use, such as cocoa, oil palm, or agricultural crops (Lowe 1977).

Although these systems appeared impractical in the face of the intensive pressure from expanding human populations, they were very desirable from an ecological point of view. Management systems that encourage natural regeneration usually are not seriously affected by disease and low soil fertility, because the naturally occurring species are adapted to local conditions (Jordan and Herrera 1981). Erosion usually is not a problem, because bare soil is seldom exposed on a large scale (Poore 1976).

A silvicultural system similar to the old shelterwood system of Africa has been tried in an evergreen seasonal forest on low-fertility soils of Suriname (de Graaf 1982, Jonkers and Schmidt 1984). Sustained-yield management of native tropical forests also has been tried within the Amazon Basin (Rankin 1985). However, because data on nutrient depletion by harvest are available for the Suriname site, that case is presented here.

Jonkers and Schmidt (1984) and de Graaf (1982) describe the system. The harvest operation starts with the felling of selected trees. The proportion felled varies, but in the example given 40% of the basal area was considered commercial. After removal of the logs, a silvicultural treatment of girdling and poisoning is initiated. This treatment is necessary to increase the growth rate of the valuable species remaining on site by reducing the competition from the nonvaluable ones and stimulating recruitment of valuable species. After treatment, the diameter increment of desired individuals is greatly increased. After about 8–10 years the growth rates decline, and followup treatments are necessary to keep growth of desired species high. A second harvest should be possible after 20 years. Total exploitable yield in the selectively harvested native forest is about 2 t/ha/yr.

Light exploitation by removal of 20 m^3 of wood from the forest was predicted to result in only a very small nutrient loss (Fig. 6.12) from an ecosystem near Kabo, Suriname (Fig. 1.2). The stocks remaining in the soil appeared adequate to prevent a nutrient deficiency.

Economic Considerations

Costs of silvicultural treatments are an investment to increase the future yields of the forest. Without the treatments, growth of commercially valuable wood is very slow. Poisoning undesirable trees increases the cost but also increases the profit after 20 years.

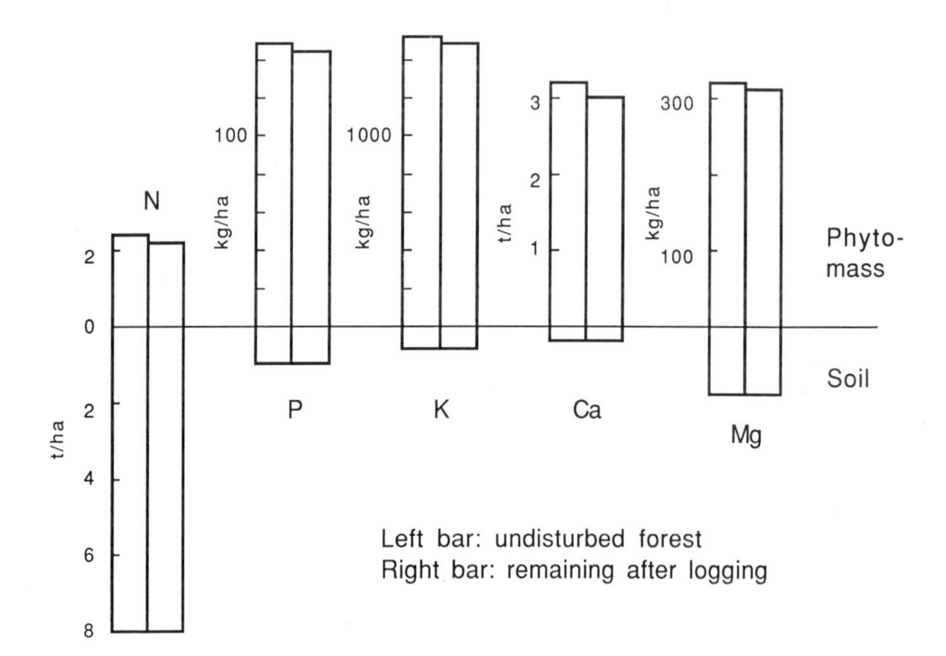

Figure 6.12. Estimated effect of removal of 20 m^3 of wood on the nutrient capital of the virgin forest at Kabo, Suriname. (Graph drawn from data in Jonkers and Schmidt 1984)

Despite the potential for this system, de Graaf (personal communication) is very pessimistic about its adoption in Amazonia as long as uncut virgin timber remains. Sustained-yield management requires a long-term investment. This investment discounted over the investment period lowers the net profit when the timber is sold. In contrast, owners of virgin stands of timber can sell their lumber for less because they have no such investment costs. Only when the cheaper timber from virgin stands disappears will the managed timber be salable at a profit.

7. Plantation Forestry*

Case Study No. 9: The Jari Project, Pará, Brazil

Charles E. Russell

In the 1950s, Daniel K. Ludwig, one of the world's richest men and owner of numerous international corporations, anticipated a global shortage of wood fiber for pulp. To meet this anticipated shortage, he and his advisors began looking for a site that would be suitable for development of a large-scale pulp plantation (*Time* 1979, Kinkead 1981). Eventually, they settled on a site in the state of Pará, in the Amazon Basin of Brazil, and purchased 4650 square miles of land near the Jari River (Fig. 7.1) at a price of about $3 million (*Time* 1976).

An important reason for the selection of the Jari site may have been the feasibility of obtaining a large contiguous tract of land at a low price per unit of land area (Kinkead 1981). The possibility of tax-free imports of machinery and loan guarantees also may have had an influence in the selection of Brazil as the host country (Gall 1979). The reasons given by the Executive Director of Jari for the selection of the site were: (1) an apparently stable government that was interested in investment by a foreign organization, (2) a large, uninterrupted block of land, (3) a moist or wet tropical climate, and (4) a deep-water port (Posey 1980). Perhaps the most important reason for selecting a site in the

---

* Except where otherwise cited, descriptions, data, and findings in this chapter are from: Russell, C. E., 1983. *Nutrient Cycling and Productivity of Native and Plantation Forests at Jari Florestal, Pará, Brazil*. Ph.D. dissertation, Institute of Ecology, University of Georgia, Athens, Georgia, USA.

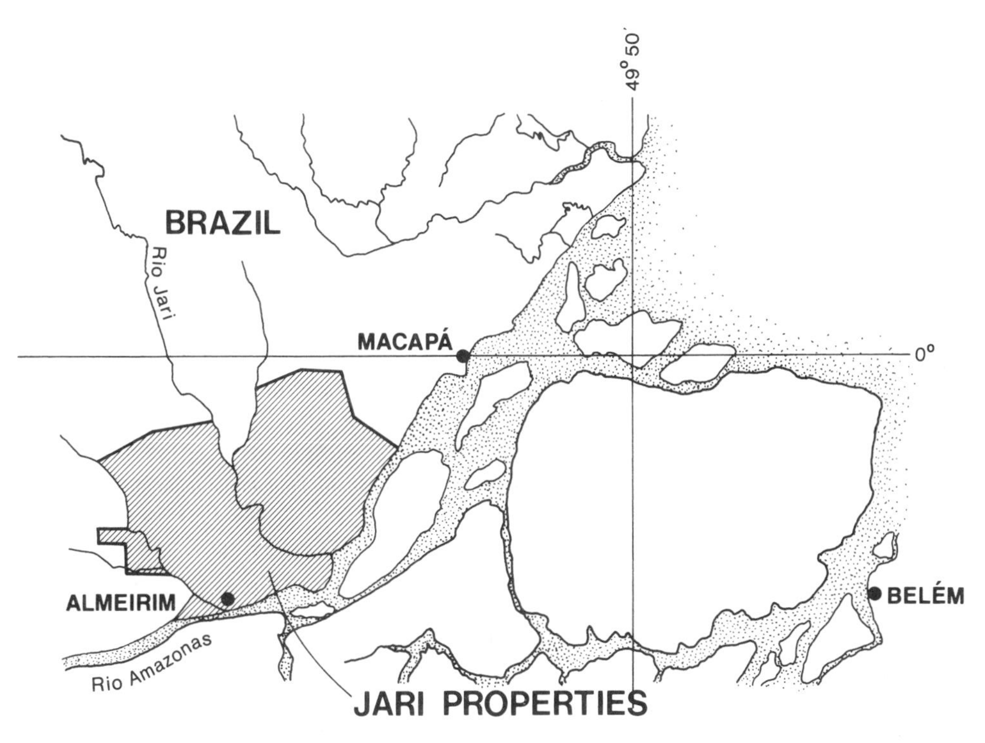

Figure 7.1. The properties of Jari Florestal e Agropecuária, approximately 350 km west of Belém, Pará, Brazil.

Amazon was Ludwig's belief that the Amazon region had a great and unlimited potential for high and sustained yield of commercial crops (Masello 1979).

The objective of the Jari enterprise was not only to raise trees for pulp but to develop an entire integrated pulp production project that included a mill suitable for producing high-quality pulp ready for delivery to overseas markets. To accomplish this goal, a massive development scheme was undertaken. A port was built on the Jari River. A railroad system with several diesel locomotives was set up to transport pulpwood from the distant regions of the plantation to the central mill location. Hundreds of kilometers of roads were built to facilitate truck transport of wood from the field sites to the railroad loading yards. A complete town, called Monte Dourado, was built, including schools, medical care units, a supermarket, bank, churches, individual houses for technical and supervisory personnel, dormitories for field labor, a central feeding facility, motor pools, machine shops, and a fuel depot (*Time* 1979). A series of outlying centers, called silvavilas, also was built to house and feed workers in remote areas of the plantation. Air service, complete with airport and planes, was established to ease the problem of slow access to and from Belém, 350 km miles to the east and the closest port of access. The single most expensive item was the pulp mill itself, built in Japan for $400 million and floated across two oceans and up the Amazon River (*Time* 1979). By 1981, the total investment in the Jari project was approximately $1 billion (Kinkead 1981).

Other projects also were developed within the framework of Jari. Sawmills were built to utilize some of the native trees cleared from the plantation plots. A factory was constructed to process the large deposits of kaolin mined on the Jari property east of the Jari river. A major rice farming operation was estabIsihed on *varzea* (floodplain) in the southern part of Jari along the Amazon River. Cattle and water buffalo for beef also were raised as part of the Jari operation (Fearnside and Rankin 1982, Hartshorn 1979).

The most significant part of the Jari project has been the pulp operation. After a search of several years for a pulp species that could grow rapidly in the wet tropics, Ludwig's advisors recommended *Gmelina arborea* (Kinkead 1981). It is a species that often establishes readily in plantations, has early vigorous growth, and sprouts readily to enable a rapid start to second and subsequent rotations. The tree produces a good quality, short-fiber pulp and the wood seasons readily, is easily pulped, and has satisfactory peeling qualities (Greaves 1979). Because of the success of this species in plantations in Africa and Asia, Ludwig had trials carried out in Costa Rica and Panama, where the trees are purported to have grown a foot a month (Kinkead 1981). Based on these results, Ludwig decided to use this species on the Jari plantation.

Large-scale plantings of *Gmelina* commenced in 1969 on sites that were cleared by bulldozers. Slash was moved into long rows (windrows) and burned. The *Gmelina* is reported to have grown well along the windrows but to have failed in other areas (Greaves 1979). By 1976, clearing of the original forest by heavy machinery had been abandoned (*Time* 1976). The reasons were: (1) it removed topsoil, (2) it compacted the topsoil, and (3) it was expensive (Posey 1980). Heavy equipment was replaced by laborers with axes and chainsaws. Despite the evidence for faster growth of trees on sites in which the slash and topsoil is left in place, clearing and slash removal was resumed again in the early 1980s (Hartshorn 1981). It is termed "complete site preparation," and the rationale is that removal of stumps and logs permits mechanized brush control during the first few years of the plantation when naturally occurring species compete with plantation species (Hartshorn 1981).

Despite the experimentation with site preparation methods, the production of *Gmelina* at Jari was 40% below projected rates (Kinkead 1981), and as a result trials with other species began. Many of the new sites were planted with *Pinus caribaea*, var *hondurensis*, and *Eucalyptus deglupta* (Posey 1980). Many of the poorest *Gmelina* plantation were ripped out and replanted with these other species (Kinkead 1981).

In 1980, an opportunity arose to evaluate the possibility that low nutrient stocks of the soil and nutrient removal during harvest of trees could be factors involved in the low plantation productivity. Specifically, the objective of the study was to determine changes in nutrient dynamics and primary productivity when the native forest is cut and converted to plantation, during the course of plantation growth, following the harvest of the first tree crop, and during the beginning of the second rotation.

Site Description

The Jari property consists of more than 1 million ha on both sides of the Rio Jari, the last major tributary on the north side of the Rio Amazonas (Fig. 7.1). The annual rainfall is about 2300 mm. The driest months are October and November, and the wettest are April and May. The topography is rolling with differences of several hundred meters between higher plateaus and valley bottoms.

Soils in the northern half of the plantation are generally higher in clay content than those in the southern half. Sites for this study were restricted to the southern half of Jari. The predominant soil type there is a sandy Ultisol (described as a ferralic Arenosol by Chijicke, 1980), with a sandy surface layer over a clay loam subsoil. It is acid (pH around 4.0) and low in potassium, calcium, and available phosphorus.

Study sites were selected on the low, rolling Ultisols. Although species composition of the native forest appeared to vary, even where soils and topography were similar, the biomass of the forest as judged by number, diameter, and height of trees appeared to be relatively uniform. The average top of the canopy is about 40 m, with occasional emergents. The size-class distribution of canopy and understory trees as determined by the point-quarter method (Cottam and Curtis 1956) in 264 quarters is given in Figure 7.2.

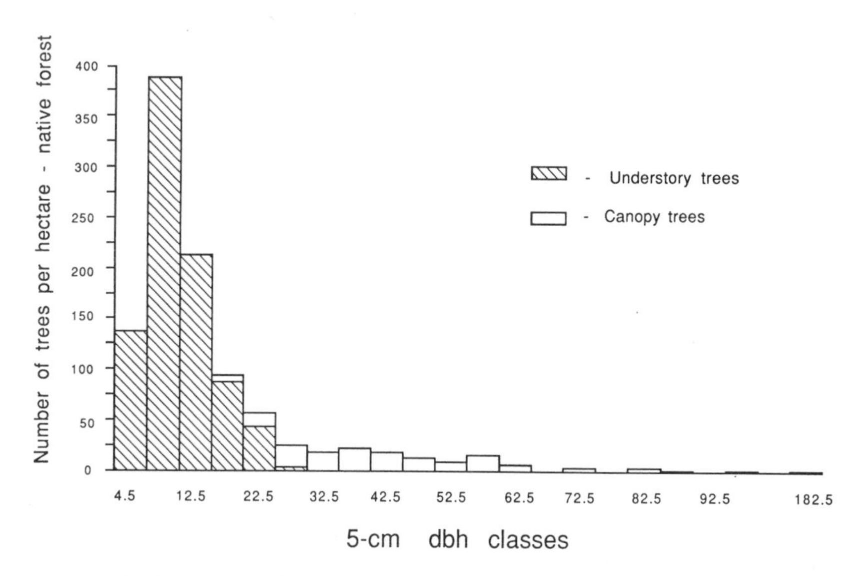

Figure 7.2. The frequency distribution by 5-cm dbh size class of 528 native forest trees studied at Jari.

Methods

Three 10 × 10 meter quadrats were established in each of the following types; (1) native forest, (2) a recently established pine plantation (6 months old), (3) a 9.5-year-old pine plantation, (4) an 8.5-year-old *Gmelina* plantation, and (5) a 6-month-old pine plantation established on a site that had supported *Gmelina* for 8.5 years. Locations of the sites within the Jari property are shown in Figure 7.3. It was assumed that the plantation sites, before harvest, all had had stocks of biomass and nutrients similar to that of the native forest site.

Each of the plots was harvested completely and dry weight of all aboveground components was determined. Vegetation and soils to below rooting depth were sampled for nutrient analysis. Biomass variability was high between the plots of native forest, and another method was used to improve the estimate of native forest biomass. Allometric equations to predict tree biomass, including roots, were determined from whole trees, exhumed with the use of backhoes. These equations were used with the point-quarter data (Fig. 7.2) to estimate biomass over a large sampling area.

Nutrient input for each of the sites was determined for 1 year with rainfall gauges and rainwater collectors. Nutrient loss was the concentration of soil drainage water determined with soil water collectors (lysimeters) times volume of soil water flow from calculations based on Molion's (1975) climatonomic study of the energy and moisture fluxes in the Amazon.

Average yearly wood production of the plantation forests over the life of the plantation was the stock of biomass divided by the age of the plantation.

Results and Discussion

Biomass Stocks and Productivity

Stocks of biomass for the 8.5-year-old stand of *Gmelina,* the 9.5-year-old stand of pine, and the native forest are shown in Figure 7.4. If standing dead *Gmelina* are acceptable for pulp, then this *Gmelina* plantation produced about 108 tons of wood in 8.5 years, or 12.8 t/ha/yr. If only the living trees are used, then the average yearly production was 9.9 t/ha/yr.

The best growth of *Gmelina* at Jari was reported to be close to 14 t/ha/yr (Woessner 1982). However, the average productivity in a large sample of plots throughout Jari was much lower. A plot of stemwood biomass stocks and biomass accumulation as a function of stand age (Fig. 7.5) shows rates lower than 6 t/ha/yr. Hornick et al. (1984) carried out calculations of *Gmelina* productivity based on data also supplied by Jari management and reported an average annual volume increment of 13 m^3/ha/yr. Because the specific wood density of *Gmelina* at Jari averaged 0.38 (Russell 1983), 13 m^3 would be equivalent to 4.9 t dry weight biomass. This growth rate is considerably below rates for *Gmelina* in other regions of the world and is also below average for other plantation species (Wadsworth 1983).

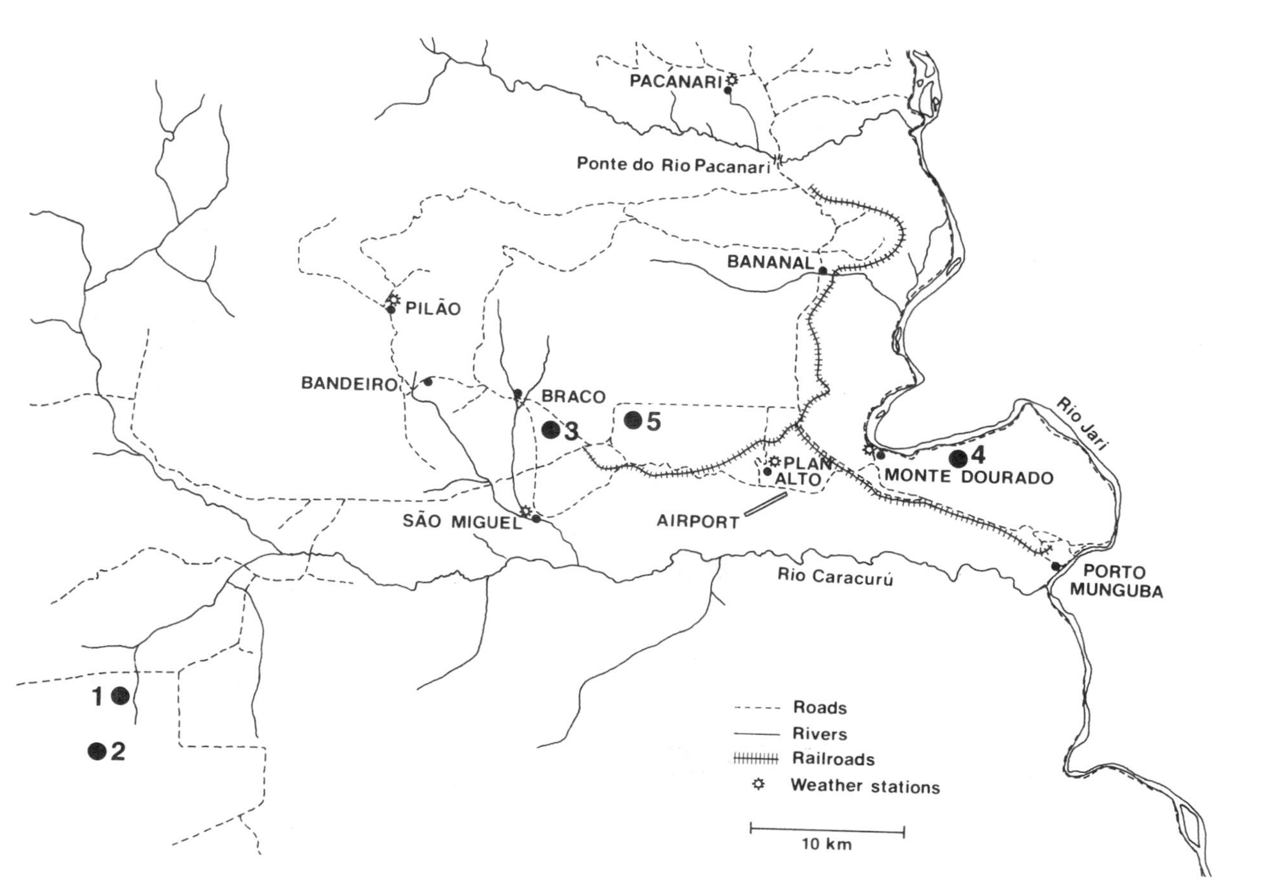

Figure 7.3. Map showing location of study sites within the Jari properties. 1, Native forest; 2, newly planted pine; 3, 8.5-year *Gmelina;* 4, 9.5-year pine; 5, first crop *Gmelina* converted to second crop pine. Local grid blocks are 10 km on a side.

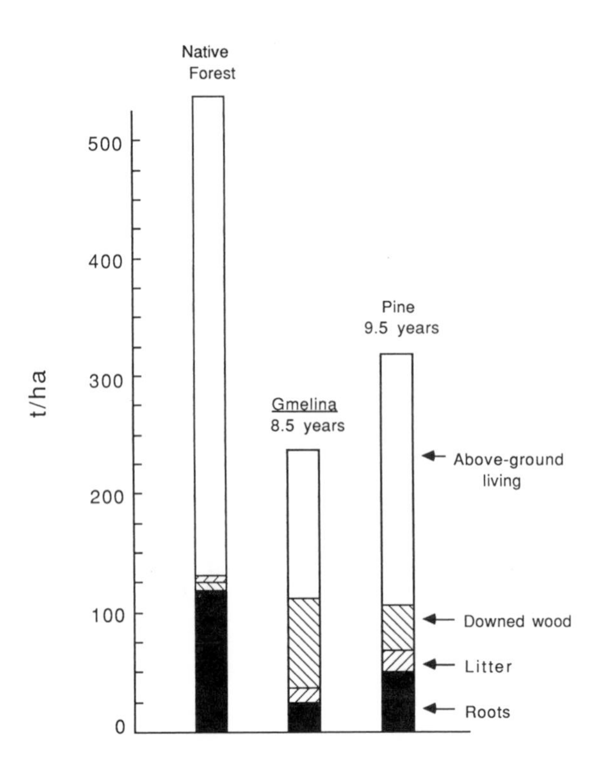

Figure 7.4. Biomass of the 8.5-year-old stand of *Gmelina,* the 9.5-year-old stand of pine, and the native forest.

Dividing stemwood biomass of the 9.5-year-old pine plantation (Fig. 7.4) by age results in an average annual production throughout the life of the stand of 16.9 t/ha/yr. However, as in the case of *Gmelina,* averages throughout the Jari plantation were lower and came to less than 6 t/ha/yr after 7 years (Fig. 7.6). The mean annual biomass increment of *Pinus caribaea* in nine studies cited by Wadsworth (1983) from other parts of the world ranged from 8 to 23 t/ha/yr.

Nutrient Dynamics

Nutrient dynamics during the conversion of native forest to pine plantation and during the growth of pine plantation to 9.5 years are summarized in Figure 7.7. This figure shows stocks of nutrients in each of the major ecosystem compartments during conversion of native forest to plantation through harvest of the first crop. Time zero on the horizontal axis is the time at which the forest was cut and burned. To the left of time zero are the stocks of nutrients in the undisturbed native forest. Compartments are stacked, that is, the stocks of nutrients in each compartment are added to the sum of those plotted below, so that the top point of the graph at any time represents the sum of stocks of nutrients in the entire ecosystem.

Most of the calcium and potassium in the native forest ecosystem is incorporated in living biomass or litter (96% and 87%, respectively). When the native forest was cut during site preparation activities prior to plantation establish-

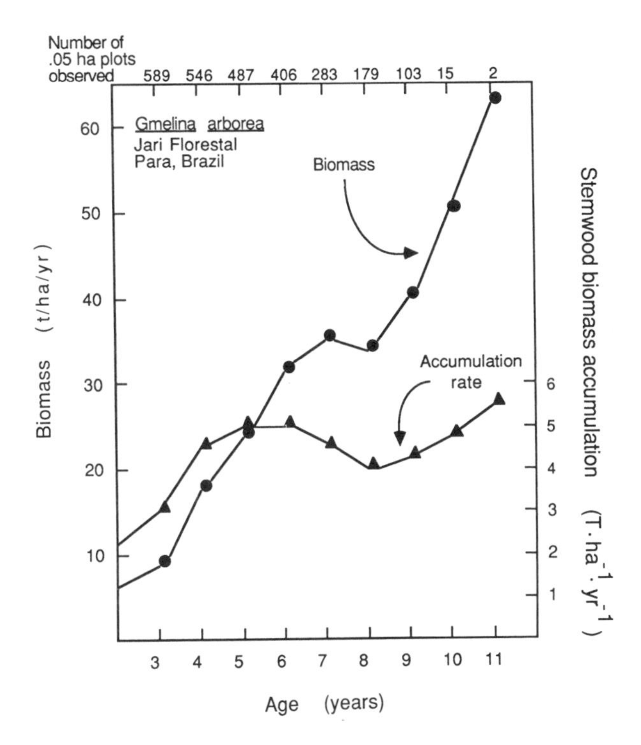

Figure 7.5. Stemwood biomass stocks and biomass accumulation rate of *Gmelina arborea* as a function of stand age. The number of 0.5-ha plots on which the data points were based are indicated along the top of the graph. (Figure prepared by R. Schmidt, Chief, Forest Service, Department of Natural Resources, Commonwealth of Puerto Rico, based on data supplied by the Inventory and Control Department of Jari.)

ment, trunks of the larger trees frequently were removed; some were used as fuel in the pulp mill, while attempts were made to use others as saw timber. In some sites, all the large woody biomass was removed before the remaining slash (branches, stumps, leaves) was burned. Figure 7.7 represents a site where all large woody biomass was removed, and the notation "losses to harvest" at time zero shows that a considerable proportion of the total ecosystem stocks of calcium and potassium are removed as a consequence of removal of the logs. Lower proportions of phosphorus and nitrogen are lost because a greater proportion of these nutrients are in the soil.

Burning of the remaining slash and litter results in the sharp increase in calcium, potassium, and phosphorus in the soil because of downward movement of ash from the burned biomass. Nitrogen in the soil shows a decrease following the burn because of volatilization of this element by the fire. Not all of the slash is burned, however, as shown by the compartment labeled "decomposing slash." This compartment gradually becomes smaller throughout the 9.5 years of the first rotation, until at the time of pine harvest most of the remains of the original primary forest have disappeared.

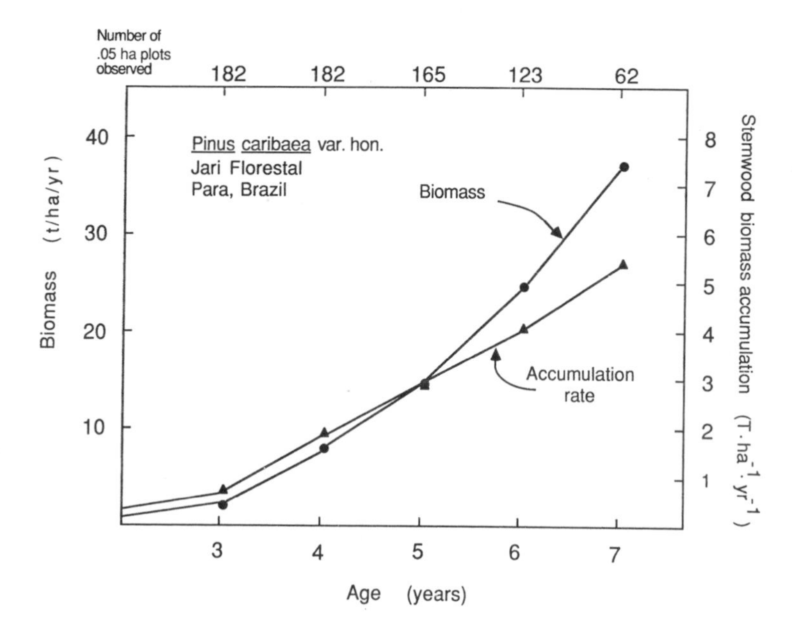

Figure 7.6. Stemwood biomass stocks and biomass accumulation rate of *Pinus caribaea* as a function of stand age. The number of .05-ha plots on which the data points were based are indicated along the top of the graph. (Figure prepared by R. Schmidt, based on data supplied by the Inventory and Control Department of Jari.)

Net leaching losses occurred only during the first few years of plantation growth. To simplify the figures, net leaching losses are plotted as though they occurred all at once at the time of the burn.

The calcium, potassium, and phosphorus in the litter layer decrease as a result of the burn but gradually build up during the course of plantation growth. Nitrogen in litter is not plotted because of the small proportion in this compartment.

During the growth of the pine plantation, stocks of calcium, potassium, and phosphorus increase in the pine biomass compartment. Part of the increase comes from the nutrients released from decomposing slash, part comes from soil stocks, which diminish throughout the course of the rotation, and part comes from atmospheric deposition. These three sources can account for the increases in calcium and potassium in the pine biomass, but they cannot account for the relatively large increase in phosphorus. The increase in phosphorus probably comes from phosphorus stocks in the soil that are held in insoluble form in iron and aluminum compounds. Soil phosphorus plotted in

---

Figure 7.7. Nutrient stocks in primary forest (to the left of time 0, horizontal axis) and ▶ changes in stocks during cutting and burning, cultivation, and harvesting of pine plantation.

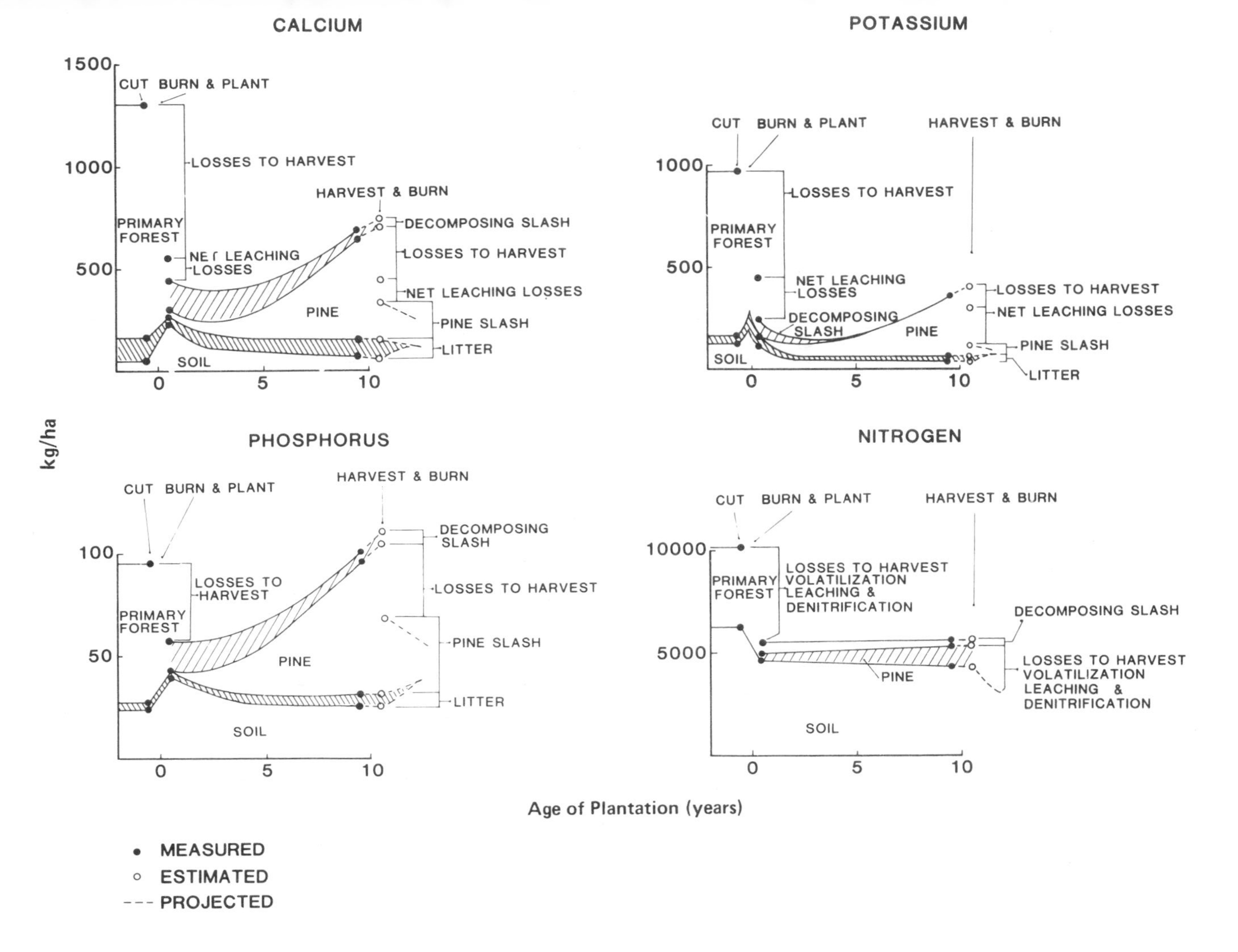
CALCIUM
POTASSIUM
PHOSPHORUS
NITROGEN
CUT
BURN & PLANT
LOSSES TO HARVEST
HARVEST & BURN
PRIMARY FOREST
NET LEACHING LOSSES
DECOMPOSING SLASH
PINE SLASH
LITTER
PINE
SOIL
LOSSES TO HARVEST VOLATILIZATION LEACHING & DENITRIFICATION
1500
1000
500
100
50
10000
5000
0
5
10
kg/ha
Age of Plantation (years)
MEASURED
ESTIMATED
PROJECTED

Figure 7.7 is soluble phosphorus, but soluble phosphorus is only about 15% of total phosphorus (determined by total digest) in surface soils, and less than 1% of total at depths below 30 cm. Total phosphorus in the soil probably is several thousand kilograms per hectare, and the increase in phosphorus stocks in the pine biomass may come from this pool of insoluble phosphorus. The exact mechanism of phosphorus transfer between pools is not clear but may involve mycorrhizae, chelation, pH change, or microbial metabolism fueled by carbon compounds from the roots and litter.

Close to half of the nitrogen in the ecosystem is lost by the harvest and burning of the primary forest. Only a small proportion of the nitrogen in the soil is incorporated into the pine biomass.

Following harvest of the first rotation of pine, there appears to be enough calcium in the pine slash, litter, and soil for another crop of pines. Phosphorus also appears not to limit stand development. Nitrogen, although at a level of about half of the primary forest, still seems to be sufficient for another crop, although the stepwise decline, if continued for several rotations, could lead to a nitrogen limitation. Potassium seems to be the critical nutrient. Removal of the pine trees will result in an elimination of almost all the potassium in the ecosystem. Stocks in the soil, litter, and pine slash at the time of harvest of the first rotation are considerably lower than stocks in soil, litter, and slash at the beginning of the first rotation. Potassium appears to be the most critical nutrient at the beginning of the second rotation.

The management sequence of primary forest to *Gmelina* to pine (Fig. 7.8) results in nutrient dynamics that are similar, with one exception, to those in the primary forest to pine conversion. *Gmelina,* as well as pine, seems to be able to mobilize the insoluble phosphorus in the soil, and so this nutrient is probably not limiting to growth. Nitrogen may be a long-term problem in *Gmelina* as well as in pine. Whereas there is an increase in total stocks during growth of the *Gmelina,* the increase is small, and nitrogen may limit long-term exploitation. As in the pine plantation, potassium appears to be the nutrient that will become most immediately limiting.

The one nutrient that is strikingly different in behavior in the *Gmelina* plantation is calcium. There is a dramatic increase in calcium in the ecosystem during growth of the *Gmelina*. To try to discover the source of this calcium, soil samples were reanalyzed, this time using complete acid digestion. Results (Jordan, unpublished data, 1983) showed that the calcium that is extractable or exchangeable with the standard dilute double acid method represents only about 50% of total calcium in the upper soil, and 10% or less in the lower soil horizons. The additional calcium obtained by the acid digest may represent calcium bound in recalcitrant humic compounds that is not dissolved by the standard extraction solution. In contrast to calcium, analysis for potassium in

Figure 7.8. Nutrient stocks in primary forest (to the left of time 0, horizontal axis) and changes in stocks during cutting and burning, planting, and harvesting of *Gmelina* and planting of pine. ▶

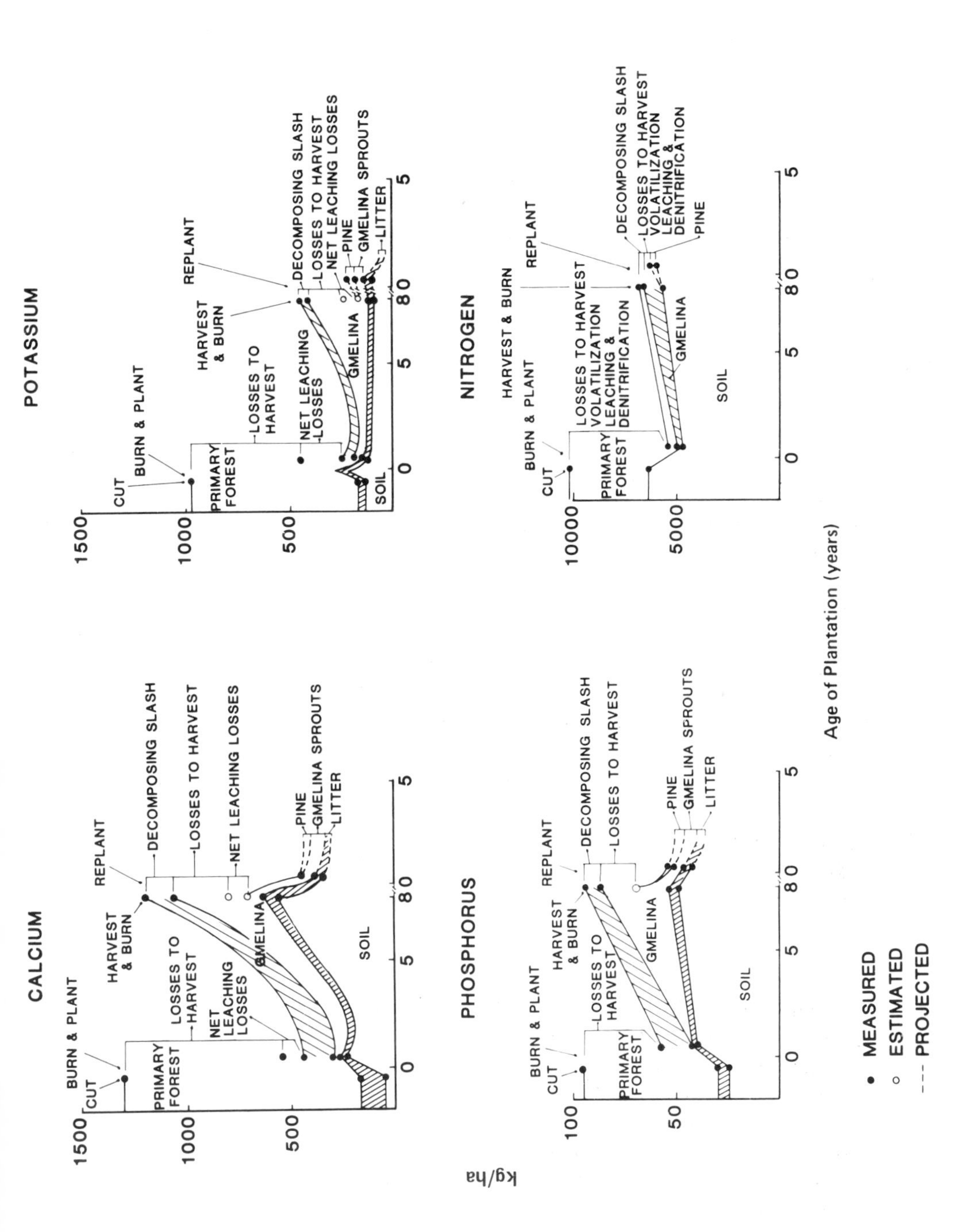
CALCIUM
POTASSIUM
PHOSPHORUS
NITROGEN
CUT
BURN & PLANT
PRIMARY FOREST
LOSSES TO HARVEST
NET LEACHING LOSSES
HARVEST & BURN
REPLANT
DECOMPOSING SLASH
GMELINA
PINE
GMELINA SPROUTS
LITTER
SOIL
VOLATILIZATION
LEACHING &
DENITRIFICATION
kg/ha
Age of Plantation (years)
MEASURED
ESTIMATED
PROJECTED

the total soil digest showed no additional potassium compared to dilute acid extractable levels.

Nutrients and Productivity

Were low levels of soil nutrients at Jari responsible for the relatively low rates of first rotation productivity? The data in Figures 7.7 and 7.8 show nutrient losses that could lower second rotation productivity below first rotation levels. However these data are not sufficient to conclude that nutrients have been responsible for the low first rotation productivity.

Evidence for possible importance of nutrients in the first rotation comes from an adjacent pair of 10.8-year-old pine plantations. Both plots were identical, except that in one the topsoil and charred slash was removed with a bulldozer to facilitate planting of seedlings. Bulldozing removes about 58% of the potassium, 80% of the calcium, 67% of the magnesium, 38% of the phosphorus, and 28% of the nitrogen on a plot.

Average total biomass, including leaves, roots, and bark of 12 trees, 10.8 years old, harvested on the bulldozed plot was 99.5 kg, whereas on the non-bulldozed plot, seven harvested trees averaged 278 kg each. Trees on the plot cleared by bulldozer had only 36% of the biomass of trees where the topsoil and slash was left in place. A photograph comparing the two plots has been published by Jordan (1985b).

Although soil compaction was not measured in the two plots, a higher bulk density in the bulldozed plot could have been another factor causing the lower productivity in that plot. Compaction inhibits root growth and lowers water storage capacity.

Other environmental problems besides soil fertility could have played a role in the low productivity of *Gmelina* and pine at Jari. There have been occasional outbreaks of insects and disease. A fungus, *Ceratocystis fimbriata,* apparently has affected growth of *Gmelina* plantations. Even on the best soils at Jari, *Ceratocystis* has caused significant economic losses (Fearnside and Rankin 1985). Leaf cutter ants (*Atta* sp.) also have frequently caused problems (Fearnside and Rankin 1980).

Fertilization

Fertilization could be a remedy for low productivity caused by nutrient deficiencies. However, the cost of fertilization should be balanced against the increase in revenues resulting from the faster growth of the fertilized stands. Calculation of this balance is difficult because of the continually changing costs of fertilizers and prices of pulpwood. Accounting decisions, such as assumptions about depreciation of capital equipment also play a role in the decision. Fearnside and Rankin (1982) estimated that the cost of fertilization at Jari would be about 6% of the gross income available when the pulp was sold, but they were not able to predict whether fertilization would be profitable because they did not know what proportion of the gross income was net profit.

Christianson (1984) calculated that with no fallow to reestablish soil fertility, an average rate of fertilization of 402 kg/ha/yr would be required at Jari. Hornick et al. (1984) reported that current practice is to fertilize *Eucalyptus* at the time of planting with 249 kg/ha, and that some of the pine and *Gmelina* sites are also fertilized.

Sale of Jari

In early 1982, a majority interest of the pulp and mining operation was sold to a consortium of 27 Brazilian firms for $280 million (Fearnside and Rankin 1982). Ludwig's associates said that failing health of the 84-year-old Ludwig was one reason for his giving up on Jari, but others were of the opinion that the billionare quit because Jari had ceased to be a good prospect. "Pesky government regulations and a xenophobic attitude among Brazilians further discouraged him" (*Time* 1982). Hartshorn (1981), in his analysis of the decision by Ludwig to sell Jari, stated:

> Company officials attribute Mr. Ludwig's decision to Brazilian government "red tape" and a failure to fulfill promises made to Mr. Ludwig. Non-company observers point out that Mr. Ludwig's investment in Jari far exceeds earlier plans and that profligate spending and poor management have contributed to the difficult financial position of the Jari operation. [p. 19:1]

Official reasons for the failure have not included environmental problems, such as low soil fertility and disease. The reluctance to suggest that these factors may have been important is curious, but this is not the first example of such reluctance. Well before production figures from Jari were available, Meggers (1971) wrote: "The persistence of the myth of boundless productivity in spite of the ignominous failure of every large-scale effort to develop the region constitutes one of the most remarkable paradoxes of our time."

Jari has continued to operate since the sale, and its financial balance still is of interest. The June 1985 issue of the *International Society of Tropical Foresters News* cited Jari as being among the top 100 agribusiness firms. The article stated that for 1983, it had reported a $2.1 million (US) operating profit. This amounts to a return of 0.2% on the initial investment of $1 billion. On the $280 million Brazilian investment, it is a return of 0.8%.

High economic return on an investment is not the only criterion of success for a development project. Social and political factors also can be significant, as they were in the case of the Trans-Amazon Highway development (Chapter 6). However, the primary objective of Jari clearly was return on economic investment. In this respect, success appears to have been limited.

8. Large-Scale Development in Eastern Amazonia*

Case Study No. 10: Pasture Management and Environmental Effects Near Paragominas, Pará

Robert J. Buschbacher, Christopher Uhl, and E.A.S. Serrão

Background

In the early 1960s, Brazil was facing severe economic problems, especially in agriculture. The problem was caused in part by Brazilian economic policies. There was little investment in agriculture, import tariffs increased costs of agricultural machinery and chemicals, and high export taxes and overvalued exchange rates made Brazilian farm products costly on the international market. Social as well as economic problems also affected agricultural productivity. Tenant farmers and sharecroppers were often not able to gain access to land. Formerly accessible lands in the southern states of Paraná and Rio Grande do Sul were closed to settlers, and this further reduced agricultural options for peasant farmers. As a result, there was an exodus of population from rural areas (Davis 1977, Foweraker 1981, Hecht 1984, Wood and Wilson 1984).

Economic reform was a major factor behind the military coup of April 1964, which replaced the regime of Brazilian President Joao Goulart with a group of generals. One of the steps taken by the new government was the establishment

* Vegetation data from Buschbacher, R.J.,C. Uhl, and E.A.S. Serrão, 1984. Forest Development Following Pasture Use in the North of Pará, Brazil. Paper presented at the First Symposium on Development in the Humid Tropics, EMBRAPA, Belém, Brazil, November 1984. Nutrient data from Buschbacher, R.J., unpublished.

of a Superintendency for the Development of the Amazon (SUDAM). Its mission was to stimulate agricultural development in the Amazon, a region perceived to be rich in developmental potential. One of the steps taken was to initiate a tax incentives program to promote more corporate investment in the Amazon Basin. The purpose of the SUDAM tax incentives program was to mobilize companies in São Paulo and other parts of Brazil to reinvest their taxable incomes in projects in the Amazon (Foweraker 1981).

The plans for development in the Amazon region had various motivations, such as opportunities for landless peasants and national security. However, the most important concern seemed to be integration of the Amazon region into the national economy of Brazil. This national integration implied greater economic linkages of the Amazon hinterlands to urban centers, facilitated by development of infrastructure and the creation of investment credits. The idea was that the developed south of Brazil was the achievable future of Brazil's backlands. Through this "manifest destiny," the new agricultural frontier in the Amazon was to provide a solution to vital economic questions and thus to help legitimize the governmental regime (Hecht 1984).

Achievement of economic sustainability was an important goal of the development plans. The government recognized that substantial investment was required to increase the infrastructure to the point where agriculture would become profitable. Once this was accomplished, the hope was that profits would stimulate further investments, especially from the private sector, and the system would become economically sustainable. To accomplish this, legislation was enacted that exempted economically important companies already in the Amazon by 1974 from all taxes for a period of 10 years. A second provision granted companies a 50% reduction on their corporate income taxes earned in other parts of Brazil if they reinvested taxable income in the Amazon. A third part of the legislation provided that the purchase of all farm machinery would be exempt from import taxes and duties (Davis 1977).

Cattle ranching seemed to be a direction for development that promised relatively easy profits. Compared with the other agricultural options in the region, such as pepper, cacao, and rubber plantations, ranching seemed relatively easy to initiate and maintain (Hecht 1984). Sudam hoped that over 500 large cattle ranches would be established under this program in the Amazon and central regions of Brazil. One of the first companies to take advantage of the new program was the well-known King Ranch of Texas. In 1968, King Ranch, in collaboration with the combined Swift–Armour Company of Brazil, was granted authorization to establish a 180,000-acre cattle ranch near the town of Paragominas, state of Pará (Davis 1977), and soon many other ranches also became established in the area (Foweraker 1981) (Fig. 8.1).

Environmental Considerations

Shortly after large-scale pasture development began in the eastern Amazon, the increasing conversion of rain forest to pasture was sharply criticized by environmental scientists. Because of the very low fertility of the soils, maximum

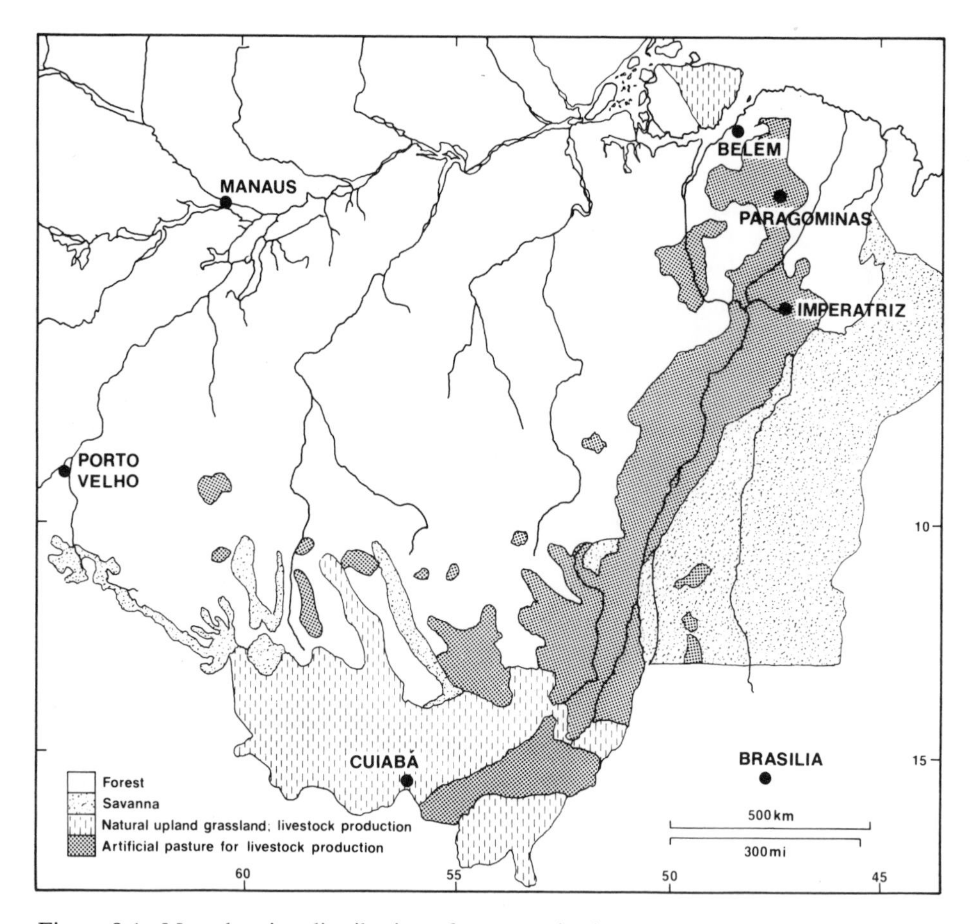

Figure 8.1. Map showing distribution of pastures in the eastern Amazon region in the late 1970s. (Adapted from Hecht unpublished dissertation)

carrying capacity the first year following conversion of forest to pasture was predicted to be only about 0.4 head of cattle per hectare. Rapid loss of nutrients by leaching and fixation would cause the maximum possible stocking rate to decline to half that by the third year (Fearnside 1979).

Because of the low carrying capacity, overgrazing was anticipated. Overgrazing would result in trampling and compaction of the soils, and erosion would carry away topsoil and nutrients. Weeds, many unpalatable and poisonous and much better adapted to degraded soils, would quickly become established and outcompete the pasture grasses. Because of these possibilities, conversion of forest to pasture was ranked least desirable of all types of development activities in Amazon rain forest areas (Goodland 1980).

The fears of ecologists about environmental degradation by grazing for the most part became reality. By the middle to late 1970s, pasture degradation was widespread. As a result of one to two decades of various combinations of clearing, burning, grazing, weeding, and mechanical scraping, many pastures in the Paragominas region of the eastern Amazon appeared to be in very poor condition. At the same time, fiscal incentives for pastures were being curtailed (Uhl and Buschbacher 1985). By 1978, about 85% of the ranches in Paragominas had failed, according to the Director of the Pará State Cattlemen's Cooperative (Hecht 1984).

By the early 1980s, many of the pastures had been abandoned for several years, and some of them were degraded to the point that questions arose about whether a forest could ever become reestablished on the sites, or whether there was a permanent conversion to heathlike vegetation. To evaluate the effect of cattle grazing on nutrient stocks, productivity, and ability of forest to regenerate, we undertook a study near Paragominas in pastures subjected to varying degrees of exploitation.

Types of Pasture Management

A preliminary survey of ranches in the region showed that pasture formation and maintenance practices varied widely. Because pasture use history could greatly influence succession following abandonment, it was important to distinguish the types of pasture management and use.

1. Low intensity. Following cutting and burning of the forest the area is seeded to pasture grasses, but the grasses do not establish well. Failure of grass to establish may result from conditions that are too wet or too dry at the time of seeding; a poor burn, which results in competition with pasture grass by stump sprouts; poor quality seed; or other factors. When grass establishment is poor the rancher usually does not invest in weeding. Only a few cattle are grazed. Trees become reestablished, and burning the pasture to kill them is difficult. Abandonment usually occurs within 4 years of pasture formation. Approximately 20% of the abandoned pastures in the Paragominas region fit this description.
2. Medium intensity. Following forest cutting and burning, the area is seeded to pasture grasses, which establish well. The pasture is weeded by workers with machetes, and every few years, the site is burned. Burning eliminates woody seedlings and favors grass species. Further, fire appears to improve both the productivity and palatibility of the grasses favored for pasture. This may come about through removal of dry grass litter, allowing light to penetrate to the soil, and through conversion of the litter to ash, which raises soil pH and improves nutrient availability. Grazing pressure is intermediate, about one animal per hectare. Abandonment occurs 6–8 years after pasture formation. About 70% of the abandoned pastures in the region fit this description.
3. High intensity. Following forest cutting and burning, the area is seeded to pasture grasses, which establish well. After several weeding and burning treatments, the areas are cleared with heavy machinery. This is to remove woody species that have become established despite fire and to remove

remaining logs and debris of the original forest. Thereafter, the sites are mowed mechanically each year and then burned. Grazing pressure is heavy, with more than two head per hectare. Abandonment occurs after 10 or more years of use. Sites subjected to such heavy use are relatively rare and represent perhaps only 10% of abandoned pastures in the region.

Methods

Fifteen abandoned pastures were located, and their management histories were documented by interviews with ranch owners and managers. The sites ranged from those that had low-intensity use to sites that had been heavily grazed, burned, and weeded several times and later bulldozed to remove woody residue. Time since abandonment ranged from 2.5 to 10 years. Three sites were selected for detailed study, one of which represented light-intensity use, one moderate-intensity use, and one high-intensity use.

The last date of burning or weeding was taken to be the time of abandonment from management and the earliest time at which forest regeneration could begin. All three sites were last burned or weeded 8 years prior to the date of the study. In the low intensity-use site there had been poor seed germination and grass establishement after the forest had been cut and burned. After 4 years of light grazing, the site was abandoned without weeding or reburning. The medium-intensity site had been used for about 7 years after pasture establishment, and during that time was heavily grazed, with weeding and burning about every second year. The high-intensity use site also had been a well-established pasture and had been grazed for 8 years with biennial weeding and burning. After 8 years, the site was cleared of all vegetation and woody residue by a tracked bulldozer, and replanted with grass. It was then grazed for three additional years before abandonment.

All sites selected for study were located on yellowish red soils, high in clay content. They have been classified as Latosols in the Brazilian classification system and probably are Oxisols in the United States system.

Vegetation of each site was sampled in 8 to 10 plots located at 50- to 100-m intervals along a transect across the area. Each plot consisted of a set of nested quadrats for sampling different sized vegetation. Trees in each plot were identified and counted, and biomass was estimated through the use of regression equations determined from whole tree harvests made outside the plots. Vines, herbs, shrubs, and grasses were harvested by complete plot harvest, and dry weight was determined. Fine root biomass (less than 2-mm diameter) was determined from cores, and coarse root biomass from root pits. Biomass subsamples and soil samples were taken for nutrient analysis.

Results and Discussion

Total above-ground biomass in the low-intensity site was 88.9 t/ha (Fig. 8.2), and biomass, including roots, was 95.2 t/ha. This value is approximately 25% of

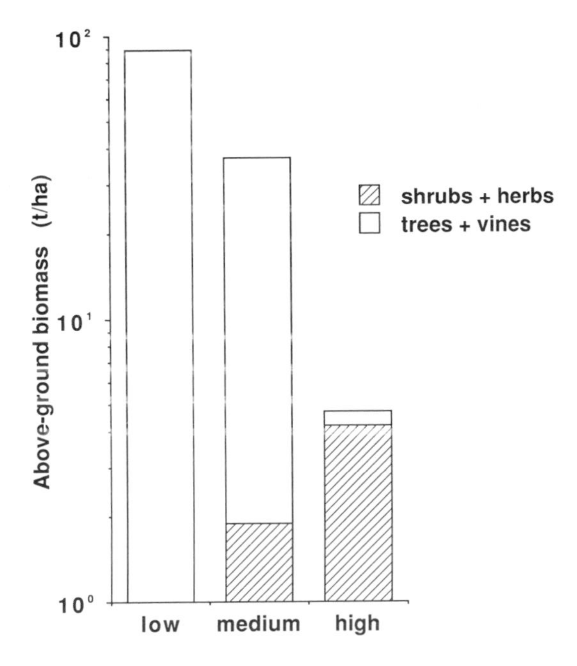

Figure 8.2. Above-ground dry-weight biomass, by life form, 8 years after abandonment of pasture sites having had low-, moderate-, and high-intensity use. Values plotted on a log scale in order to clearly depict differences in biomass of shrubs and herbs between the medium- and high-intensity use sites.

the forest biomass that might have been expected in the primary forest prior to cutting for pasture establishment. Recovery of one-fourth the original biomass in 8 years suggests that a rapid rate of structural recovery is occurring. It does not necessarily mean that full recovery will occur in 32 years, but the fact that almost all the vegetation is trees, as opposed to shrubs, herbs, and grasses (Fig. 8.2), suggests no important obstacle to reestablishment of forest.

In the site with medium-intensity use, total biomass was 45.9 t/ha, and biomass above ground was 37.0 t/ha. Shrubs, herbs, and grasses made up about 5% of total biomass, and vines made up about 16% (Fig. 8.2). The greatest proportion of the biomass is in trees and, as in the case of the low-intensity site, the moderate-intensity site should recover to forest, although not as rapidly as the low-intensity site.

In contrast to the low- and medium-intensity use sites, the high-intensity use site was almost completely dominated by shrubs, herbs, and grasses. Trees comprised 4% of the biomass. The potential for this site to recover to forest appears to be much less than for the other sites. In the 8 years since abandonment, only 5.9 t/ha of total and 4.7 t/ha of above-ground biomass had accumulated on the site. At the time of the survey, most of the plants had already reached the maximum size attainable.

Further evidence that the low-intensity site is likely to recover to forest, whereas the high-intensity site probably will not, comes from the data on species occupying the sites. Importance values for short-lived successional tree species as a proportion of total number of trees (Fig. 8.3) shows a difference of 43% versus 100% dominance, between low- and high-intensity use sites. The most common of the short-lived successional species dominating the high-use

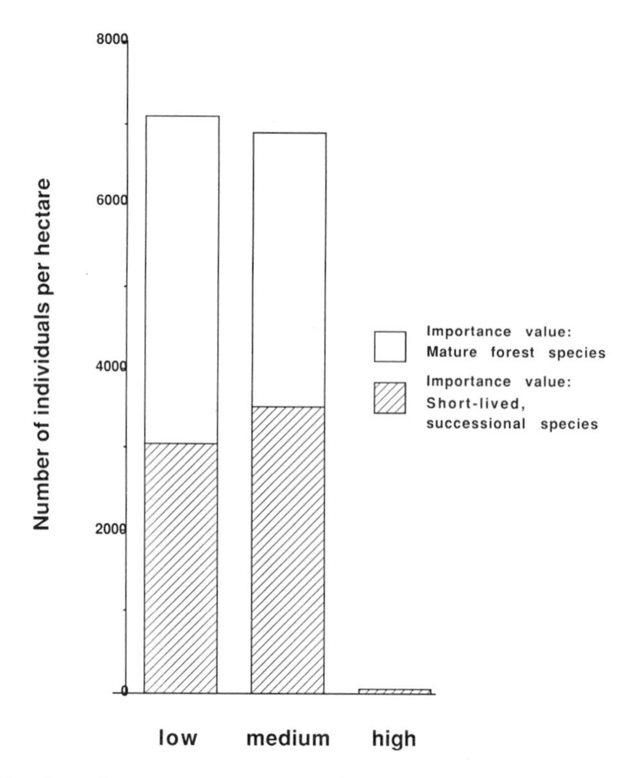

Figure 8.3. Number of trees per hectare and importance value of trees in pasture sites having had low-, moderate-, and high-intensity use. The total height of each bar is the total number of trees greater than 2 m height in each plot. Bars are proportioned between mature forest species and short-lived successional species. Proportion is based on importance value (frequency, density, basal area) rather than on number alone, to better reflect functional importance of the two groups.

site was *Solanum grandiflora,* a plant adapted to inhabit nutrient-impoverished soils on sites that are intolerable to most species. In contrast, the low-intensity disturbance site was less dominated by any one species, had a higher species diversity, and had a higher proportion of mature-forest species present. The most important genus on this site, *Cecropia,* is adapted to rapid colonization of lightly disturbed sites, such as tree fall gaps and slash and burn agricultural clearings, and has been taken as an indicator of relatively favorable site conditions (Uhl and Jordan 1984).

Nutrient Stocks

Nutrient stocks in living vegetation and soil in the abandoned pastures are given in Figure 8.4. Comparison with stocks in nearby or predisturbance primary forests would indicate the amount of restocking necessary for full recovery of the forest, but such data were not available. To suggest what stocks might have existed, comparisons are made with data from the primary forest at

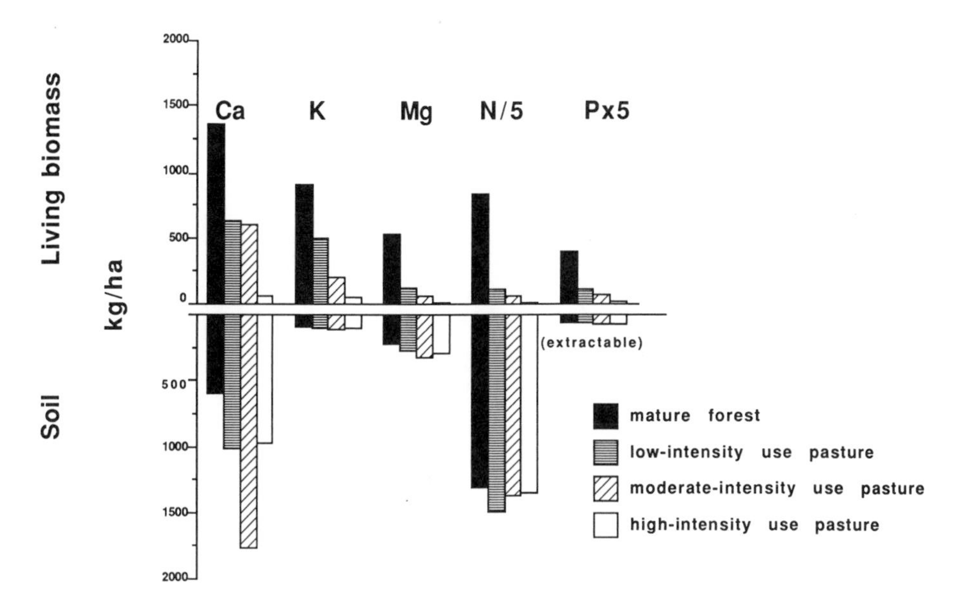

Figure 8.4. Stocks of major nutrient elements in the living biomass and soil of mature forest, and abandoned pasture following three intensities of use. Soil data for mature forest are from undisturbed forest adjacent to pasture plots. Biomass stocks are from Figure 7.4, which includes data from undisturbed mature forest in northern Pará.

Jari in the northern part of Pará. These data (from Fig. 7.4) are plotted at the left in the group of bars for each element in Figure 8.4.

The first result to examine is differences in stocks of nutrients in the living biomass. For pasture sites, the low-intensity use pasture always has the highest stocks of nutrients. However, differences between pasture sites are small compared to the difference between pasture sites and primary forest. If the Jari data for the primary forest resemble at all the stocks that might occur near Paragominas, there is a large loss of all nutrients in the living biomass following conversion of forest to pasture.

Stocks of nutrients in the soil contrast with the pattern in the biomass. In the soil, stocks are higher in the pastures than in the primary forest, although in all cases except calcium the difference is small. This pattern appears similar to the results of Falesi (1976), who claimed that conversion of forest to pasture improves soil properties, particularly calcium. However, Figure 8.4 shows clearly that a small increase in nutrient stocks in the soil comes at the expense of a large decrease in nutrient stocks in the biomass. It is the decomposing logs and other slash on the soil surface that replenishes nutrients lost from the soil through leaching and other processes.

Conversion of forest to pasture increases nutrient stocks in the soil even when fertilizers are not used. However, conversion of forest to pasture drastically decreases the total ecosystem stocks of most nutrients. It is these total stocks that must be rebuilt for forest to become reestablished on the pastures.

Rebuilding must take place from weathering of parent materials and atmospheric input. Because parent rocks are already highly weathered and low in nutrients, rebuilding stocks may be a slow process.

Factors Influencing Forest Recovery

The biomass and diversity data suggest that the abandoned low-intensity pasture site may be progressing toward forest, whereas the high-intensity site may not. Because there is little difference in soil nutrient stocks between the low- and high-intensity use sites (Fig. 8.4), other factors besides nutrients may be influencing the early course of succession.

Some of the factors that may be important, and that should be included in future studies include:

1. Compaction of the soil, which may inhibit establishment of tree species more than that of herbaceous and shrubby species.
2. A life history pattern that favors trees that sprout, when sprouts are still viable, as in low-intensity use pastures, but that favors weeds with short life cycles when disturbance is continuous, as in high-intensity use pastures.
3. Competition for growing space, which lessens opportunities for tree species to establish once the soil is thickly covered with herbs and grasses.
4. Differences in microclimate resulting from shading effects of slash, such as old trunks in lightly used pastures. Some tree seedlings may be very sensitive to the high water stress present in direct sunlight.
5. Efficiency of seed predators in locating new tree seeds in areas with lots of slash as opposed to in bulldozed pastures, which are more uniform.
6. Efficiency of herbivores in finding leaves of tree seedlings in uniform, bulldozed pastures.
7. Presence of the appropriate species of mycorrhizae for establishment of tree species. Tree symbionts may be scarce in the high-intensity use pasture.
8. Presence of appropriate species of small mammals and birds, necessary vectors for carrying seeds of many tree species.

Once tree species are established on an abandoned pasture site, stocks of nutrients available in the soil may be important in governing the rate of forest growth. Abandoned slash and burn sites in the upper Rio Negro region required over a century to rebuild stocks of nutrients equivalent to those in the predisturbance forest (Chapters 2 and 3).

Evaluation

Lack of economic sustainability of the pastures near Paragominas does not mean that the effort was not profitable for some of the entrepreneurs. Even as pasture productivity declined, the value of pasture land increased because of the infrastructure development near Paragominas, the mineral resource potential, and the generous fiscal incentives. The creation of a land market in Amazonia where previously none had existed generated an extraordinary speculative boom throughout the 1960s and 1970s (Hecht 1984). The exchange value of the land itself was far higher than the value of the beef that it could produce. Entrepreneurs depended on the increase in price of the land resulting from speculation for profits, not on the annual productivity of the land.

Increase in the value of the land was caused in part by development of infrastructure by the government. One of the infrastructure developments in the region was a road system. The roads could be used to haul hardwood species of the native forest to sawmills in Paragominas. Sawing native trees for lumber gave the wood a value higher than simply that of the ash that fertilized the soil during pasture establishment. As financial incentives for clearing of forest to establish pasture decreased, more and more land owners abandoned such efforts and concentrated on harvesting trees that would bring high prices at the mills of Paragominas. This appears to be a more rational use of the timber resources, but there are important negative aspects to the practice. We have described the damage that can result from such harvesting as follows:

> Taken alone, selective forest cutting is not a severe disturbance or cause for concern, but in the Paragominas region, several factors interact to make timber harvesting activities much more detrimental than they first appear. Since cutting is being done by people who do not own the land, there is no interest in doing the work carefully. Many more trees are cut than are actually harvested. During skidding operations when boles with good shape are dragged to spur roads, many young saplings are crushed, and dead trees and branches are scattered about. The end result is thousands of square kilometers of cut-up forest criss-crossed with soil exposed by machinery, and laden with dead slash. The extensive canopy openings and the dry slash on the forest floor turn a normally fire-resistant ecosystem into a fire-susceptible ecosystem. Fires that are set in nearby pastures to control weeds easily spread into the logged forests. The logging roads provide a route for rapid spread of fire into the forest interior. Few forests in the Paragominas area have escaped fire damage. (Modified from Uhl and Buschbacher, 1985 pp. 267–268.)

From an environmental point of view, selective logging may be as bad or worse than pasture establishment. Although integration of Amazonia into the political network of Brazil may be an important objective of development such as that near Paragominas, it would seem important to look for approaches other than creating pastures followed by selective logging.

9. Conclusion

Comparison and Evaluation of Case Studies

The objective of this concluding chapter is to compare the individual case studies and to look for principles that can be used to guide development and management of the Amazonian forest landscapes. Comparisons are made on the basis of productivity, economic factors, and achievement of social and political goals. Principles of energy analysis and the applicability to the case studies also are reviewed.

Productivity

Almost all the case studies have dealt with the effect of disturbance on productivity and agricultural or forest yield. Disturbance is both the change that occurs to the preexisting rain forest and the application of factors of production, such as fertilizers, pesticides, labor, and machinery, necessary to obtain agricultural yield. The case studies were presented in an approximate order of increasing intensity of application of the factors of production. High intensity can mean high concentration of factors per unit land, as in the case of fertilizers in continuous crop production at Yurimaguas, Peru. High intensity can also mean high total application of factors because of the large scale of the project, as at Jari. At Jari, total intensity was high, but concentration per unit land may have been lower than at Yurimaguas.

Can any insights be gained from a comparison of productivity at various intensities of development? Although yield data were not available from all case studies, there are enough data (Table 9.1) to suggest that as both the concentration and the scale of intensity of agriculture increase, agricultural productivity increases. With development, agriculture shifts from subsistence to a production of surplus that can be sold.

Direct comparisons of the case studies are difficult because of differences in crop species. However, comparisons of yield within the same class of crops tended to show similar trends. For example, technology-intensive agriculture in Yurimaguas, Peru had higher yields than shifting cultivation in San Carlos, Venezuela. Forest yields also responded to intensity of management. Total exploitable yield in the selectively harvested native forest in Suriname was about 2 t/ha/yr. In contrast, even the lowest estimates for pulp species at Jari were two to three times that value.

Increasing pasture technology, such as aerial seeding of grasses and occasional fertilization, appeared to increase cattle productivity in the Amazon. Although no data are available on cattle producton on the pastures of Paragominas, where these practices are sometimes used (Chapter 8), there definitely was a harvestable increment in beef. In contrast, in the San Carlos pasture, where technological infrastructure was completely lacking, there was no net beef production. Cattle died faster than the survivors grew.

Although it is clear that increased subsidy through management increases production, the data in Table 9.1 do not indicate whether the higher yields are

Table 9.1. Rates of Economically Valuable Production in Three Scales of Disturbance and Development Discussed in this Book[a]

Scale	Crop	Yield (t/ha/yr)	Ref.
Small scale, low intensity, and short duration			
Shifting cultivation, San Carlos	Yuca	0.7–1.5	Chapter 2
Frontier pasture, San Carlos	Cattle	0	Chapter 5
Intermediate scale, intensity, and duration—permanent plots, continuous cultivation			
Colonists, Trans-Amazon	Rice	1.6 (s = 0.7)	Chapter 6
	Corn	1.3 (s = 0.6)	Chapter 6
"Complete" fertilization, Yurimaguas	Rice	2.0–3.5	Figure 6.11
	Corn	3.0–5.0	Figure 6.11
	Soybean	3.0–4.5	Figure 6.11
Sustained yield forestry, Suriname	Hardwood	2	Chapter 6
Large scale, intensive development			
Jari	*Gmelina*	5–13	Chapter 7
Jari	Pine	6–17	Chapter 7

[a] In the Trans-Amazon study, s is standard deviation.

sustainable. Higher yields through increased intensity of management require an increased economic subsidy. Increased intensities of management must be paid for through higher profits or through other subsidies. Are the higher yields that result from intensive management sufficient to make the system profitable? If the goals are other than profitability, are the goals sufficiently achieved so that the subsidies can continue? If not, the high productivity will not be sustainable.

Sustainability of Production

Conway (1983) has defined sustainability as the ability of a system to maintain productivity in spite of a major disturbance, such as is caused by intensive stress or a large perturbation. He defines stress as a regular, sometimes continuous, relatively small, and predictable disturbance. A perturbation is an irregular, infrequent, relatively large, and unpredictable disturbance, such as is caused by a drought or flood. Lack of sustainability may be indicated by declining productivity, or collapse may come suddenly and without warning.

One stress that appeared to be important in affecting sustainability in the Amazon case studies was nutrient loss via leaching, volatilization, and fixation in mineral soil, as well as loss by harvest. The only case study that could be considered an indefinitely sustainable system of agriculture, according to Conway's definition, is the shifting cultivation system at San Carlos. There, where population density is very low, there is ample time for the naturally occurring species to obtain nutrient input from the atmosphere and rebuild the stocks of nutrients, before a plot is cultivated again. This type of sustainability, dependent only upon natural processes, can be termed ecological sustainability. The shifting cultivation in the Gran Pajonal, Peru, appeared to be ecologically sustainable when the fallow plot was in old successional forest before reuse. In the sites where periods of fallow are only a few years, there is doubt as to whether agriculture is ecologically sustainable.

The concept of sustainability, however, is not restricted to ecological sustainability. For example, continuous cultivation of annual crops at Yurimaguas may be sustainable because lost nutrients were replaced with fertilizers. This type of agriculture can be called technologically sustainable. In technologically sustainable agriculture, productivity can be sustained by man-applied subsidies of whatever factor is limiting production. Because the factors for plant production are well known, technologically sustainable agriculture theoretically can be carried out anywhere, in the Amazon, in the desert, or on spacecraft. However, for agriculture to be meaningful for development, the technology must be economically feasible.

Profitability

The essence of economic sustainability is profit. A system that is economically profitable usually will be sustained for as long as the profit continues. Economic profitability is the perspective for comparison of case studies here.

The concept of "profit" is relevant only to activities carried out within the framework of the modern economy. Slash and burn agriculture, as described in Chapters 2 and 4, is usually carried out only for subsistence. Shifting cultivators occasionally sell or exchange some of their crops for machetes or axes, but shifting cultivation is generally outside the sphere of national economic activity and therefore cannot be evaluated in terms of profit. It is not a very desirable land use from the perspective of the usual national, economic goals.

Clearing of rain forest for pasture at San Carlos, Venezuela, described in Chapter 5, seemed to have been to display Venezuelan sovereignty in a remote area subject to claims by squatters who are not nationals. This goal also is outside the profit-oriented economic argument.

The Trans-Amazon Highway scheme was profitable for certain groups. Government personnel on the highway program received a large salary bonus for working at a hardship post, and several purchased ranches and farms along the road. Most construction companies, hotels, restaurants, storekeepers, and landowners in the preexisting towns have profited from the Trans-Amazon. Several thousand peasants have titles to land that is rapidly increasing in value (Smith 1982). Nevertheless, the competition between political and economic goals prevented successful attainment of either. For example, enough funds were allocated to move the peasants onto the land, but not enough were made available to build an infrastructure to make their farming efforts profitable. As a result, farmers throughout many regions were left saddled with heavy bank debts and no opportunity to sell much of their produce. Many colonists left the area. The end result was that by 1974, Brazilian President Geisel announced that the development of the Amazon by means of small farmers had failed (Moran 1984).

Profitability for continuous cropping systems such as that at Yurimaguas, Peru is difficult to achieve. In addition to the costs of production sustained by farms close to major markets, remote jungle projects have the additional costs of hauling fertilizer to the site and hauling products from the site to market. In places where there is a transportation and marketing infrastructure, continuous cultivation with fertilization can be profitable (Sanchez et al. 1982). The cropping system at Yurimaguas was claimed to be economically sustainable by the scientists involved (Sanchez et al. 1982), although an independent evaluation (Fearnside, unpublished, 1985) contradicted this claim.

The Agroecology project at Tome-Assú, Brazil has become very profitable and in recent years has been the third highest income-producing region in the state of Pará (de Oliveira 1983). Most of the profit comes from the sale of black pepper. However, other products, such as cocoa, vanilla, and rubber, also are sold in Belém and in southern Brazil according to scientists at the Brazilian Agricultural Agency (EMBRAPA) in Belém.

Because the experiments on management and selective harvest of the native forest are relatively new, there are no data on profitability of these sustained yield systems. The system of management being tried in Suriname is similar to the African "tropical shelterwood system" (Fox 1976). This system was abandoned because profitability was low compared to that for other uses for the land

(Lowe 1977). Management of native species in the Amazon also has an economic obstacle. Timber sales from managed plots cannot compete for profit against "liquidation sales" of unmanaged virgin timber (de Graaf personal communication, 1985). Only when unmanaged virgin timber becomes unavailable, either through depletion or through creation of parks and reserves, can sustained-yield management of native species become profitable.

The plantation forest project (Chapter 7) at Jari, Brazil, was begun in 1967. By 1981, the total investment by Daniel Ludwig and his corporation had been approximately $1 billion (Kinkead 1981). The economic profitability of Jari apparently was expected to be high, higher certainly than small-scale cultivation of the type at Yurimaguas or along the Trans-Amazon. In early 1982, a majority interest of the pulp and mining operation was sold to a consortium of 27 Brazilian firms for $280 million (Fearnside and Rankin 1982). In 1983, Jari reported a profit of $2.1 million. This is a rate of return of less than 1% on the investment. Economic sustainability of an industry with returns this low may be doubtful.

Like the Jari pulp plantation, large-scale ranching was initiated in the expectation of large profits. Large-scale pastures turned out to be economically sustainable only as long as the ranchers received some type of government subsidy. When these subsidies were withdrawn, few new pastures were cleared and existing ones were gradually abandoned (Chapter 8).

Social and Geopolitical Criteria

The Trans-Amazon Highway colonization program is a case study in which social and political goals were an important aspect of the project. Reviews of the success of these goals ranged from moderately positive to very negative. D. Miller (1983) stated that the principal objective of resettling 100,000 families from the drought-stricken northeast was not met, and the entire small-farm sector floundered under a number of constraints. The plan met more success in terms of the second major objective, that of more fully integrating the Amazon region into national Brazilian life. Part of this was a result of the establishment of a bureaucracy in the Amazon region. In fact, those who benefitted most from the Trans-Amazon highway were the urban entrepreneurs and bureaucrats.

Smith (1981) asserted that the Trans-Amazon colonization scheme failed on all three of its major goals. The first goal was to provide a safety valve for the poverty-stricken northeast, a region with 30 million inhabitants, increasing by 1 million a year. The second goal of the project was to fill a demographic void in a region occupying half of Brazil's territory but containing only 4% of the nation's population. The Brazilian government saw the highway as placing an indelible stamp of sovereignty on the resources of the region. However, through 1981, only 8000 families had been settled by the Instituto Nacional de Colonizacao e Reforma Agraria, the federal agency responsible for administering the project. The highway scheme had not even come close to achieving the colonization target and absorbed less than 1% of the population growth of the northeast, according to Smith. Wood and Wilson (1984) agreed that the number of persons absorbed by the Amazon frontier was minimal relative to the magnitude of population growth and the size of the rural to urban migration stream.

The third goal was to create access to mineral and timber reserves that would fuel the country's economic growth. However, neither lumber nor mineral operations have provided significant income for the Trans-Amazon settlers (Smith 1981).

Moran (1981) was more optimistic about the Trans-Amazon Highway project. He pointed out that it was important to judge success on an individual level, rather than to try to evaluate the whole project collectively. In his book (Moran 1981), he cited a large variability in the success of colonizers at a village near Altamira. He found that about 40% succeeded by making adjustments to their new environment, whereas the rest reverted to a state of dependency.

Sawyer (1984) concluded that the Amazon frontier should not be seen as a solution for national or international problems of either an economic or a demographic nature. He suggested that capitalist agriculture should only be stimulated in a few small areas and for a few crops that have a real chance of success, in order to avoid waste of public resources and widespread damage to the environment.

It is almost impossible to judge the other case studies by political and social criteria. Shifting cultivation does not solve any pressing problems for the government and so would presumably be undesirable, although it might be desirable from the viewpoint of the indigenous peoples. The pulp plantation at Jari employed and relocated too few people to have any social impact on a nationwide scale. Large-scale pastures in the state of Pará may be having an important social impact, but a judgement would be premature.

Judgements on the social and geopolitical desirability of development projects frequently are made only from the perspective of the national government in a capital far removed from the project. If the desirability of a development project were judged by indigenous peoples a different set of conclusions might result. For example, such huge projects as pulp plantations and pasture encroach on the habitat of the indigenous Indians, and so from their perspective these large projects are undesirable.

Environmental Limitations

By most of the criteria used for comparison, the largest scale development projects, that is the pulp plantation and the large pastures, were less successful than had been anticipated. Although bureaucratic and logistical problems played a role, the major reason for lack of success seems to have been failure to consider environmental limitations of the Amazon Basin. What were, and still are, the limitations?

In the case of the Jari pulp plantation, disease and parasites of the trees were implicated as influencing productivity. The continuous hot, humid climate, and a genetically uniform host resulted in an exponential growth of pest species.

Weedy vines growing on trees in the Jari plantation also could have decreased production (personal observation, 1981). Weed competition appeared to play a major role in large-scale pastures, where shrubs and unpalatable herbs replaced valuable pasture grasses.

However, perhaps the most important environmental limitation in the Amazon Basin, and the factor that appears to have influenced productivity and its sustainability in almost every case study, is low soil fertility. In almost all cases, low and declining productivity was correlated with nutrient loss, through fixation, as in the case of phosphorus; through leaching, as for potassium; or through losses through harvest, as for calcium and potassium in the pine forests at Jari. Nitrogen loss through volatilization may have been important in the large pastures because of repeated fires. Nitrogen loss frequently is important during burning of tropical grassland (Nye and Greenland 1960).

Correlation of low yields with low fertility or high nutrient loss does not, of course, prove that these factors are the cause of the low yields. Correlation merely indicates that these factors are statistically related. The only case in which there is statistically significant proof that nutrients are critical is the continuously cultivated plots at Yurimaguas, Peru (Fig. 6.11). This graph shows that fields treated with inorganic fertilizers had a significantly higher crop yield than unfertilized fields.

Why were environmental limitations ignored by the highest levels of government in the case of Paragominas, and by the top business management in the case of Jari? The limitations were certainly recognized by scientists working in the Amazon, as well as experienced local farmers, as emphasized by Moran (1981). Large-scale development is driven by many forces on a national and international level. For example, an important goal of business decisions is to generate high profits for investors quickly. An important goal of government decisions is to respond to social pressures before the next election. Thus, government and business leaders tend to seek short-term solutions or respond to promises of short-term solutions. Environmental limitations, which may appear after a number of years, are not of particular interest in this context. If environmental limitations have a potential to become serious, the decision makers only hope is that they do not appear during their tenure in office, or before the initial profits are made. The need to respond to immediate pressures seems to be the reason for ignoring environmental limitations.

Energy Comparisons

The case studies can also be compared with other agriculture and foresty developments using energy analysis. This is a common approach to evaluation of development because it offers a method of converting a diversity of system inputs and outputs into a comparable currency, "energy" (Pimentel et al. 1973, Fluck and Baird 1980, Odum 1983). However, there have been a number of criticisms of energy analysis. One is the problem of system boundaries (Leach 1975). Conclusions depend in part upon whether the system is taken to be the field, the farm, the region, or an entire country. A second criticism is that energy is measured in calories or joules, but the quality of energy varies (Leach 1975, Hyman 1980). For example, a calorie of beef is of higher quality than a calorie of grass, because many calories of grass are needed to produce one calorie of beef. A third criticism is that energy analysis does not consider relative scarcity, in contrast to economic analysis, which does (Webb and Pearce 1975, Huettner 1976). Thus, a calorie of petroleum is a constant, regardless of the amount left in the well, but the price of that calorie goes up as the reservoir becomes depleted.

These and other criticisms of energy analysis have been recognized as limiting the range of possible comparisons. Nevertheless, the opportunity to use a common currency makes the energy analysis technique attractive within restricted systems. By using the same boundary criteria for a variety of systems, and by restricting output comparisons to the same type of crop, meaningful comparisons can be made (Pimentel 1980). Further, the criticism that energy analysis does not consider scarcity, as does conventional economic theory (Huettner 1976), may not be completely relevant because the two types of analyses accomplish different objectives (Gilliland 1975). Economic analysis is based on scarcity, and market prices react to real or perceived shortages. For example, the market will drive up the price of petroleum and insure its wise use only when exhaustion of petroleum reserves appears imminent. Market price and traditional economics often serve for efficient short-term allocation of resources. In contrast, if the objective is to encourage efficient use regardless of the amount in reserve, and to prevent squandering of a resource until such time as it becomes scarce, then energy analysis may be a more effective way of making decisions.

Trends in Energy Use

As a context for comparing and evaluating the case studies, there are a variety of analyses of energy in agriculture. For example, Pimentel (1980) carefully defined the cultural energy used in a variety of agricultural and forestry systems. Cultural energy includes labor, fuel for tractors, fertilizers, pesticides, and the farm machinery itself. The energy value of machinery and chemicals was limited to the energy necessary for synthesis in the factory. Energy costs did not extend beyond the boundaries of the factory. Using Pimentel's (1980) values, Hall (1984) showed that energy-intensive agriculture has higher crop yields than primitive systems, in terms of weight, volume, or energy of the crop yield. However, the energy use efficiency, that is, the ratio of energy output to energy input of a system, is lower in the high-intensity system. Heichel and Frink (1975) also showed that as agriculture changes from primitive systems, such as slash and burn, to systems that are energy intensive in both concentration and scale, such as the highly mechanized midwestern United States agriculture, the total energy yield increases but energy use efficiency decreases.

The general trend of increasing energy yield and decreasing energy use efficiency from preindustrial systems to full-industrial systems does not seem

Table 9.2. Efficiency of Seven Agricultural Systems[a]

Category, Agricultural System	Energy Yield, (kcal/ha/yr) ($\times 10^3$)	Energy Ratio (Output/Input)
Preindustrial		
1. New Guinea	349	14.2
2. Wiltshire 1826	1,766	12.6
Semiindustrial		
3. Ontong, Java	3,527	14.2
4. S. India 1955	10,104	10.2
5. S. India 1975	15,883	9.7
Full industrial		
6. Russian collective	1,926	1.3
7. S. England 1971	10,728	2.1

[a] From Bayliss–Smith (1982).

to be limited to a series of agricultural systems in which the crop species is always the same. The trend occurs even when the crops are of different species, as long as comparisons are restricted to a general type, such as grain farms (Cox and Atkins 1979). Only when higher trophic levels, such as beef and egg production, are included in the comparison does the trend fail to hold (Steinhart and Steinhart 1974). The reason that it does not hold in such cases is that food such as beef, which is high in protein, requires a relatively high amount of energy for synthesis, and so beef production is a relatively inefficient energy production system, even in primitive grazing areas.

An example of the trends is presented in Table 9.2. There is a general trend of increasing energy yield with increasing concentration and/or scale of agricultural intensity. The semiindustrial system of Southern India in 1975 is high, perhaps because of the improved varieties of rice used in that system. The Russian collective farm is low, possibly because of management difficulties. The trend in energy use efficiency is somewhat more consistent. There is a decline from subsistence to industrial systems. These results are similar to those of Fluck and Baird (1980), who present data on energy output as a function of energy input for about 50 agricultural systems. Increasing the energy input increases the output but decreases the efficiency of cultural energy use.

Energy Trends in the Case Studies

The general trend of increasing energy yield with increasing intensity of development also seems to hold true for some of the Amazonian case studies. Net energy analysis has been carried out for the slash and burn case at San Carlos. Uhl and Murphy (1981) found a net energy yield the first 2 years of 8,406,345 kcal/ha in the form of yuca tubers, and a labor input of 606,464 kcal, for an output/input yield ratio of 13.9 : 1 over the 2-year period. This falls within the expected range for preindustrial systems shown in Table 9.2.

Compared to San Carlos, continuous cropping at Yurimaguas, Peru (Chapter 6) should have a higher energy yield but lower energy use efficiency because of the intensive use of cultural energy in the form of fertilizers. Average yield per hectare per year of rice was 3 t and that of corn was 3.9 t at Yurimaguas (Sanchez et al. 1983). The caloric value of rice is about 2952 kcal/kg (Rutger and Grant 1980) and that of corn is about 3960 kcal/kg (Pimentel et al 1973). Total energy yield of rice at Yurimaguas then was possibly 8,856,000 kcal/ha/yr, or about twice the energy yield at San Carlos. Energy efficiency data for rice cultivation at Yurimaguas are not available, but dry season rice cultivation in the Philippines has a ratio of 3.36 (Rutger and Grant 1980), much lower than at San Carlos. Total energy yield of corn at Yurimaguas was 15,444,000 kcal/ha/yr, almost four times the yield rate of energy at San Carlos. Energy efficiency of corn at Yurimaguas is not known, but intensive cultivation of corn with fertilization almost always has an output/input ratio of less than 4.2 (Pimentel and Burgess 1980).

Energy quality becomes important in comparisons of agricultural and forestry yields. If the quality of energy is defined as the ratio of the energy output and the amount of energy required to synthesize the product (Costanza 1980, Odum 1983), grain crops should have a high quality, because the plant must be grown before the fruit can be produced. The energy quality of the grain requires the energy used by the plant for growth. In contrast, in forestry systems, the bulk of the plant itself, the wood, is the product of interest, and therefore wood should have a lower energy quality. Saying it another way, an agricultural system should have a lower quantity of high-quality energy, whereas a forestry system should have a higher quantity of lower quality energy. If this argument is true, it could change conclusions based on Table 9.1, which shows relatively high biomass yields at Jari. In terms of energy, yields might be much closer to those of the crop production systems.

Energy use in the large-scale pastures also is not comparable with agricultural land uses. Energy production in the form of beef requires passage of the energy through two trophic steps, instead of just one as with plant crops. A longer food chain results in greater energy losses because of the energy loss at each food chain transfer.

Despite the difficulty of comparing energy use efficiency in the case studies, the general principles that have emerged from the comparisons cited above from Bayliss-Smith (1982) and others are supported by the case studies. Those principles are that along an energy use gradient from primitive agriculture to industrialized agriculture, total energy yield increases but energy use efficiency (cultural energy) decreases.

Energetic Value of Nature's Services

The general principle that energy yield is high in high-input systems is intuitively reasonable. The trend that appears strange is the high energy efficiency of primitive systems compared to industrialized systems. Within industrialized systems, the efficiency of scale concept is familiar; large systems seem to use resources more efficiently than do small systems, at least up to some limit. Why do large-scale agricultural systems not show a similar efficiency of scale compared to small or primitive systems?

The reason may be that analysis of cultural energy does not include all the important energy inputs into the system. For example, it does not include the energetic values of the services performed by the native forest. In the studies analyzed above, the energetic input into the agricultural systems by the native forest was ignored. As will be shown below, the total energy input into primitive systems is probably much greater than previous analyses have suggested. The seemingly high energy use efficiency by these systems probably is caused by failure to include all energy inputs into the system.

The idea that nature's services have an energetic value has been developed by Odum (1983). Inclusion of the energetic value of the services of nature changes completely the conclusions about energy use efficiency drawn from the comparisons in Table 9.2. The apparently very favorable energy balance results from subsidies provided by nature, which are not accounted for in the energy balance calculations. For example, the services of sheep in concentrating energy during grazing may account for the seemingly high efficiency of agriculture in Wiltshire in 1826. In Java, it is the services of the naturally occurring coconut palm trees that constitute the subsidy. No planting and little cultivation of palm trees is required, and most of the energy expended is simply in picking coconuts. In the paddy fields of southern India, nitrogen fixation by naturally occurring algae in flooded fields is an important subsidy not accounted for in the energy balance.

For shifting cultivation, the energy value of the trees that are cut and burned should be included, because without the nutrients and organic matter from the forest trees, slash and burn agriculture would not be possible. The trees, throughout the period of fallow, perform a service in that they take up nutrients from a relatively unavailable form in the atmosphere and the subsoil and incorporate them in biomass. Once in the biomass, the nutrients can then become available to crop plants when the trees are cut and burned. The work of forest trees in accumulating and concentrating nutrients is an essential component of shifting cultivation.

Most studies of the energy balance of shifting cultivation do not account for the fallow value of the trees, and consequently the net energy balance of slash and burn agriculture seems high. For example, Rappaport (1971) found an energy output/input ratio of 16.5 : 1 for shifting cultivators in New Guinea. Norman (1978), using data gathered in Africa by Clark and Haswell (1970), found energy output to input ratios averaging about 18 : 1 for cereal crops. For the case study of slash and burn agriculture in Chapter 2, Uhl and Murphy (1981) calculated a ratio of 13.9 : 1 in the Rio Negro of Venezuela.

How can the energy value of the fallow service of the forest be calculated for the slash and burn agriculture at San Carlos? How do we measure energy quality? Is the fallow service of the trees equal only to the energy value of the wood? Is it simply the number of calories released when the wood is oxidized through complete combustion, or does the service of accumulating nutrients over the lifetime of the tree result in a higher quality energy? Let us make the most conservative assumption that there would be no enrichment of energy quality. In this case, the caloric value of the forest biomass is about 5.3 kcal/kg (Jordan 1971). The standing crop of forest vegetation on the plot at San Carlos is 324.6 t/ha. The total caloric value of the forest then is 1720×10^6 kcal/ha. In the net energy analysis of Uhl and Murphy, they found a total cultural input to the site of 6.06×10^5 kcal/ha. Total energy input into the cultivated site then is the cultural energy plus the caloric value of the trees. This sum equals 1720.6×10^6 kcal/ha. Adding in the most conservative estimate of the services of nature to the energy balance calculation changes the ratio from 13.9 : 1 to 0.005 : 1.

The energetic value of the trees dwarfs all other inputs necessary for shifting cultivation. When the services of the trees in mobilizing the nutrients are considered as a cost of Amazonian agriculture, the energy efficiency makes a dramatic shift from a net gain to a net loss. From this point of view, shifting cultivation is a very inefficient agricultural system.

Shifting cultivation is not the only type of agriculture in the Amazon that takes advantage of the nutrient mobilization performed by native trees. Two other important examples from the case studies are the Jari plantation and the pastures near Paragominas. In these cases, it was the nutrients released from the burned and decomposing remains of the original forest that made plantations and pasture possible. When we determine the efficiency with which the rain forest resource is used when it is cut and transformed to plantation and pasture, we must therefore use the energetic value of the native forest.

Assuming that different species of trees have approximately the same caloric concentration per gram of wood, Figure 7.4 indicates the efficiency with which the native forest resource is used to produce pulpwood at Jari. The figure suggests that the rain forest resource is not being used very efficiently at Jari. About half of the energetic value of the forest is lost in producing the first crop of trees. Further, this analysis does not include any cultural energy costs, which would make the output/input ratio even lower. However, the plantation forest itself provides a service for the succeeding crop, in that it rebuilds, at least in part, the stock of nutrients lost during the removal of the native forest (Figs. 7.7 and 7.8).

In contrast, following conversion of forest to pasture, nutrient stocks are not rebuilt but are continually depleted, except where beef production may not be the primary motivation, as at San Carlos. Conversion to pastures may be an extremely inefficient energetic use of the rain forest.

We can calculate the efficiency of energy use based on data from the case studies and from the literature. The San Carlos pasture study can be used for forage productivity data. Figure 5.6 shows that production of grass edible by cattle at San Carlos was between 1 and 2 t/ha/yr (dry weight) when the pasture was derived from mature forest. This is higher than yields estimated by Fearnside (1978) for other pastures in Amazonia, where commercial fertilizers are not used. An absolute maximum time for use of unfertilized pastures is 10 years, resulting in a total grass production of around 15 t/ha. The original forest may have a biomass somewhere between 300 t/ha (Fig. 2.1) to 500 t/ha (Fig.

7.4). Taking a midrange value of 400 t, and assuming a similar energy content of grass and wood, the output/input ratio is 15/400 = .038. This energetic efficiency of about 4% does not include the cultural energy supplements necessary for pasture.

"External" Costs of Beef

Much of the beef produced on rain forest pastures in Latin America has gone into production of hamburgers for fast food chains (Myers 1984). Because the forest is essentially free to the cattle producer, its value is not reflected in the price of the meat. In economic terms, the forest is an "externality." What is the external cost of rain forest destroyed to produce one hamburger? Fearnside (1978) calculated that it requires 2.20 kg dry weight of grass to produce 1 kg live weight of beef in Amazonia. At 4% efficiency, it would take 55 kg of forest to produce 2.20 kg grass. If the forest is 400 t/ha, 55 kg would occupy 1.38 m^2. A quarter-pound hamburger then costs 1.38/4 = .34 m^2 of rain forest. A hectare of rain forest can be converted to 29,411 quarter-pound hamburgers. A million hamburgers sold is equivalent to 34 ha destroyed when Amazon rain forest is the source of the beef.

Other Services of Nature

This book has focused on sites of the upland Oxisols and Ultisols of the Amazon *tierra firme*. Within Amazonia, there are areas of richer soil, such as the *varzeas* along the Amazon, which have a much higher productive potential and so may be more appropriate for development (Shubart 1983). This is certainly a rational suggestion, but what must be taken into account is that the higher productivity of the *varzeas* is also dependent on the free services of nature. The annual floods along the Amazon and many of its tributaries, including the main channel called Solimões above Manaus, deposit loads of nutrient-rich silt eroded from mountains to the west. This annual enrichment is largely responsible for the high productivity of the *varzeas*.

Building dams eliminates the deposition of sediments along the river banks, where it can be used by the farmers. Instead, the silt is deposited behind the dams, a service that is counterproductive. The energetic services of nature, instead of being beneficial, become destructive.

Conclusions on the Energetic Value of Nature's Services

Including the energetic services of nature in comparisons of Amazonian land use alternatives leads to a conclusion opposite to that reached when only cultural energy is considered. Systems that rely on felled and burned trees as a basis for soil fertility are energy inefficient. Cutting and burning the forest leads to an energy cost that is very high compared to the cultural inputs of most systems, such as labor and supplies. Cutting and burning a forest results in higher yields for a few years, but after that period production is no longer sustainable, and total yield derived over the lifetime of such a system is low.

The services of the trees in enriching the soil and maintaining system fertility are used up in only a few years.

The case studies of shifting cultivation exemplify the importance of the services of the native forest in reestablishing the fertility of the soil, so a plot can be cultivated again. The case study of pastures near Paragominas shows that when the native vegetation is kept out by burning and clearing, the services of nature are prohibited, and the resilience of the system may be destroyed beyond the point of recovery. Clearly, the large-scale development projects in the Amazon did not consider the importance of nature's services, especially those provided by trees in mobilizing and conserving nutrients.

In contrast, systems in which the structure of the forest is maintained are much more efficient over a long period of time. When the structure of the forest is maintained, annual yield may be lower, but yield is sustainable. The total output over the lifetime of the development project is higher, and so the efficiency of resource use becomes higher. The most energy- and resource-efficient development system in a naturally forested environment is one that maintains a basic structure and function of a forest, such as the sustained-yield forestry system in Suriname. However, a forest does not have to consist of native species to have a high energy use efficiency. Agroforestry, with overstory trees and understory perennials, such as the Tome-Assú system, also should be efficient. Although the pulp plantations at Jari had a forest structure, at least the first rotation depended heavily upon the nutrients previously accumulated by the native forest, and so Jari probably is closer in energy use efficiency (including services of nature) to shifting cultivation than to sustained yield forestry in Suriname. Understanding the services of nature clarifies why development projects in which trees are maintained seem to have more promise of sustainability in the Amazon than do ones in which trees are eliminated.

Efficiency of Scale

A previous section noted that the trend of energy use efficiency apparent in the literature appeared to contradict the efficiency of scale concept, that is, the general trend of increasing energy use efficiency with increasing scale of development. The reason, it appears, is that the energetic services of nature are an important component in production, and these services are usually ignored in the literature. When the energetic services of nature as well as cultural energy are considered, the efficiency of shifting cultivation, a small scale of development, appears to be very low.

But how about the large-scale pastures near Paragominas, and the pulp plantation at Jari? These also appeared to be energetically inefficient, despite their large scale. The crucial point is to distinguish between scale as in total amount, and scale as in amount per unit land, that is, concentration of intensity. Large projects like Jari and the pastures at Paragominas, although having a large-*scale* intensity of development, have a low *concentration* intensity of development. Lots of capital was invested in these projects, but it was spread thinly over thousands of hectares. The advantages of concentration of intensity did not exist. A lot of rain forest was destroyed, and its energetic services were mostly wasted.

In contrast, the projects in Suriname and Tome-Assú, although small compared to Jari, concentrated the cultural resources that were available on small areas of land. In terms of energy use per unit of land, these projects probably used cultural energy much more intensively than Jari or Paragominas. Thus, when speaking of efficiency of scale in terms of land development, it may be more accurate to talk about increasing efficiency of energy use with increasing *concentration* of energy use, rather than with increasing *scale,* that is, size, of development.

Implications for Development

If there is a single lesson to be learned from the comparison of the Amazon case studies, it is this: to be successful over the long term, development plans must couple economic and social needs with an explicit recognition of the locally variable environmental limitations of the region.

Can this be done in the nutrient-stressed environment of the Amazon rain forest? An example of development in another environmentally stressed region of the world suggests that it can. A large part of Israel is a desert, yet many communal farms have developed profitable and sustainable agriculture under these harsh environmental conditions. However, this development was carried out by first recognizing that water was a limiting factor and then designing systems around that limitation. Of course, these modern farmers use sophisiticated technologies and are supported by an extensive research program, special market opportunities, and a unique social situation. In a similar manner, development in the Amazon must begin by recognizing that nutrients will be limiting and then designing projects that will both overcome this limitation and be profitable at the same time.

However, there is another important difference between the situations in Israel and in the Amazon. In Israel, there are no "virgin lands," in the sense of the virgin lands in the remaining Amazon forest, where a resource is available only for the taking. In Israel, agriculturalists are forced to use intensively and carefully the few resources that are available.

In the Amazon, people are not forced to take such care. There is still a lot of virgin land. When virgin land that can be exploited cheaply is still available, people will not invest in managing and sustaining lands already under cultivation. As long as free resources are available, investment in sustained-yield management will be ignored. In the case of the Amazon, as long as virgin forest is available and accessible, management programs that require thought, investment, and time will usually not be practiced, except perhaps as part of a scientific experiment. It requires less effort and money, and much less time to go into the virgin Amazon forest and extract trees for charcoal, railroad ties, or lumber than it does to plant or manage a plot as part of a sustained yield forestry operation. The investment of time and money required for sustainable yield will not come about as long as it is possible to harvest the virgin forest.

The same problem exists for development of sustainable agriculture. As long as it is possible to burn the virgin forest as a source of nutrients for fertilizer, no one will spend the money for fertilizers, or the time and effort for agroforestry systems, except under unusual conditions, such as exist at Tome-Assú or Yurimaguas. In the virgin forest, all that has to be done is cut and burn. In sustained-yield agriculture, the costs of management, labor, and such components as fertilizers, herbicides, and tractors must be discounted into the future and added to the initial costs of cutting the forest. Almost anyone whose work is guided by economic profitability will choose to exploit the virgin forest. To the exploiter it is a free resource, whereas the resources necessary for sustained-yield management are costly.

Restricting Access to Free Resources

There are two solutions to the problem of accessible virgin forest making sound forest management impractical. One is to quickly cut down all the remaining virgin forest. While this may be acceptable to some, it is not to many others, for reasons soon to be presented. The other solution is to restrict access to the virgin forests. Only when virgin forest is no longer available will people be forced to apply intensive efforts to cultivate areas previously used and abandoned. When easily accessible and cheap timber is no longer available, sustained yield forestry will then be possible.

The way to achieve better agricultural practices in areas already deforested is the same as for better forestry practices. Restrict access to virgin lands. When free fertilizer in the form of burned trees is no longer available, farmers will then turn to agriculture that uses commercial fertilizers, or organic matter from agroforestry systems.

Promoting sustainable land management through restricting access to virgin lands is not difficult to do. The most important step is to stop road building into primary rain forest. It is hard work for a colonist to penetrate virgin rain forest. Rarely does significant deforestation occur as a result of a pioneer trekking into the wilderness and hacking out a homestead, as in North American pioneer folklore. Only when roads are built is there a significant influx of colonists and other developers. Building of roads in the Amazon rain forest requires heavy machinery, bulldozers, graders, bridge building equipment, and capital available only to the government or large corporations. Decisions to restrict the use of such equipment can be made at high levels of government and business.

The World Bank and the Inter-American Bank have long had loan policies that encourage road building in remote Amazon forests (Rich 1985). An understanding of the importance of restricting access to the virgin forest reveals that these policies are counterproductive to development of sustainable management practices in the Amazon. Road building opens up previously unaccessible free resources. With these free resources available, no one, or very few at least, will invest the resources in a management program that inevitably will be less profitable. Only by eliminating road building will the peasants, entrepreneurs, and others in the Amazon be forced to adopt sustainable management practices. Continued road building only pushes further into the wilderness the frontier style of exploitation, that is, cutting, using for a few years, and then abandoning for a new frontier. Although there are increasing efforts to persuade the World Bank to discontinue loans that promote environmentally destructive activities, little change in policies has been apparent through 1985 (Holden 1986).

The argument that access to lands be restricted should not be interpreted as antidevelopment. To recommend that countries of the Amazon Basin lock up the remainder of the Amazon forest is unrealistic. There is an urgent political, social, and economic need for the resources of the Amazon. The argument offered here is that new lands should not be opened up until lands already available are well used and managed. The path to wise and sustainable management of lands already opened up is to restrict access to new lands. Only when accessible lands are used wisely and well should there be further investment in opening up the frontier.

Parks and Reserves

Another way to restrict access to virgin timber is to create reserves. Brazil now recognizies various conservation units, such as national parks, biological reserves, and national forests. De Melo Carvalho (1984) reviews park development activities in Brazilian Amazonia. It is important that parks not be simply leftover areas on lands marginal for agricultural use. Parks can serve as refugia for the tremendous diversity of species in the Amazon. However, to serve as effective sanctuaries, reserves must be designed with the distributions and territories of populations as a consideration (Prance 1982).

Creating reserves, of course, does not guarantee that settlers will be kept out. Squatters, poachers, miners, and others are problems in many Amazonian parks. One solution to this type of problem was worked out by a scientific research group with a large preserve in Costa Rica. They hired the first subsistence farmer that appeared looking for land to be a park guard. Because salaried employment is frequently preferred over subsistence agriculture, the man took the job and apparently has done well in keeping out others. Along with the recognition of the needs of local people, it is also essential to create among them a recognition of the values obtained by preserving the forest in parks.

Management Possibilities

If it is possible to substantially restrict further access to virgin Amazon forest, what management options are available for *terra firme* lands now available but little or underused? In areas near major population centers, such as Belém and Manaus, and where a transportation network is developed, fruit and vegetable crops are a possibility. The market accessibility provides profits as an incentive either to fertilize or to use intensive labor, as at Tome-Assú.

A more difficult problem occurs in remote areas, where importation of fertilizers is expensive, living conditions are difficult and unattractive, and markets for products are distant. What is the best system for such regions?

The best systems are those that mimic, as much as possible, the naturally occuring forest, in order to take advantage of "nature's free services." Sustained-yield forestry, as in the Suriname case study, is environmentally very desirable. A variety of management approaches for primary and secondary tropical forests have been described by Synnot and Kemp (1976) and Kyrklund (1976), among others.

Line Planting

One management system to increase the economic value of the yield of Amazonian forests without converting them into monoculture plantations is the use of "line planting." This is a compromise between plantation forestry and management of native forest that combines some of the best elements of both systems. Line planting involves planting of valuable species along swaths cleared in the native forest. This is an advantage over management of native forests, where it is not always possible to get the most desirable species in adequate densities to sustain a harvest operation. It is also better than plantation forests because it maintains a large proportion of the native forest, which in turn conserves nutrients. The distance between rows of individuals in line plantings also lessens the problem of disease spread common in monoculture plantations. Practical experience (M. Dourojeanni, personal communication 1985) has shown that swaths in line plantings must be sufficiently wide to allow ample light to reach the seedlings. In the short term, line planting may be more expensive than monoculture plantations because large machinery cannot be brought in. In the long term, these higher initial costs should be more than compensated for by lower costs of maintenance, and sustainability.

Line plantings can be made more economical by combining them with strip harvesting of the forest. When the forest is cut in strips, dangers of nutrient leaching and soil erosion are lessened, and in addition, the newly planted trees in harvested strips benefit from leaf litter and other organic matter falling from the adjacent trees (Jordan 1982).

Perennial Tree Crops

Another way to take advantage of the nutrient mobilization performed by trees is to plant perennial tree crops, such as cacao and rubber, two crops that have been developed and are profitable in some areas of the Amazon. During harvest, only a small proportion of the living biomass is removed, and consequently, nutrient losses are relatively small. Soil is not disturbed, so compaction and erosion are minimized. Perhaps most important, the trees remain intact. A mature tree is able to extract nutrients from a greater depth in the soil.

Further, some trees fix nitrogen, such as species of the genus *Erythrina,* which are frequently planted as overstory for cacao. Many rain forest trees have epiphyllae, such as algae and lichens, that scavenge nutrients from the rainfall (Jordan et al. 1980), a phenomenon first recognized by early studies of radioactive fallout from atmospheric nuclear tests (Whicker and Schultz 1982). Another advantage of maintaining trees is that both the root litter and the leaf litter serve as energy and nutrient sources for organisms whose activity maintains favorable chemical and physical conditions in the soil. However, the market for rubber and for cacao and other fruits of tropical trees is limited, and these crops cannot be prescribed as the only solution for Amazon-wide development.

Reclamation of Degraded Areas

Another possibility is to experiment with innovative approaches to agriculture using already disturbed areas. One of the most promising ideas deals with utilization of the hundreds of square kilometers of abandoned and degraded pastures in Amazonia. Neither crops nor trees can grow well in these sites, because of low fertility and severe weed competition. Compounding the problem are the periodic fires that spread through these grasslands during dry seasons. Currently, the only way to open these sites is to clear them with heavy machinery. This, of course, degrades the soil even more, is expensive, and results in even greater need for soil preparation and fertilizer. An alternative would be the use of a system that is similar to "no-till" or "minimum-till" agriculture. In this system, herbicides are used to control the dense grass and shrubs, and disturbance of the soil with heavy plows and cultivators is minimized. This preserves the soil structure and takes advantage of nutrient conservation resulting from the presence of a large below-ground community of microflora and microfauna (Lowrance et al. 1984). This approach to agriculture is becoming increasingly more common in the United States and often is called "conservation tillage" (Gebhardt et al. 1985).

Relatively short-lived herbicides are being developed that can kill the unwanted vegetation and then decompose within less than a year, before the residues are carried into drainage streams. Once the mature grass and shrubs are dead, the undisturbed soil and decomposing organic matter on top of the soil can benefit newly planted trees or crops.

An added benefit of restoring these sites is that a road network usually already penetrates the areas.

An important research question for such sites is whether it is possible to plant crops or tree seedlings in the residue of herbicide-treated plots without further cultivation for weed control. Under certain conditions weeds might be a problem, but because they would have become established at about the same time as the seedlings control should be easier. For some types of seedlings, such as those of primary forest species, it might be desirable to have a cover of secondary tree species to provide a more suitable habitat.

When used under properly controlled conditions, herbicides can be useful. Pelletized forms are now available, and this lessens the risk to the person applying them. Dosage is much easier to control. For large areas, pelletized herbicides can be applied with precision by air.

Another Approach to Conservation

White (1967), in his well-known article "The Historical Roots of our Ecologic Crisis," developed the point that much of the present-day environmental destruction and species extinctions, at least in the Christian world, result from a religious belief that God's will is that man exploit nature for man's own benefit. For example, Genesis 1 : 26 says:

> And God said, Let us make man in our image, after our likeness, and let them have dominion over the fish of the sea, and over the fowl of the air, and over the cattle, and over all the earth, and over every creeping thing that creepeth upon the earth.

In this anthropocentric framework, man's only ethical and moral duty is toward human life. Nature has no reason for existence save to serve man (White 1967).

In an effort to slow the current rate of species loss and habitat depletion, some modern ecologists have argued that species and habitat preservation are ethical and moral acts, on the basis that species and ecosystem destruction results in a loss of resources that may have economic, as well as spiritual, value to man. For example, Westman (1977) asked "How Much are Nature's Services Worth" and pointed out the value of ecosystems to man through such services as soil conservation and nitrogen fixation. Myers (1983), in his "A Wealth of Wild Species," catalogs the potentials that tropical plant and animal species have for agriculture, medicine, and industry. Pearsall (1984) emphasizes "in absentia" benefits of nature preserves, such as the psychological benefits of simply knowing that a wild species exists. A curious aspect of these arguments is that despite the moral and ethical overtones of species preservation, the arguments still are based on a utilitarian premise. The major improvement in these arguments over the belief that man has the right to immediately exploit all he encounters in nature for his own short-term benefit is that the arguments take into consideration long-term benefits and use of the species for future generations. The arguments are still anthropocentric.

Rolston (1985), in his article "Duties to Endangered Species" discusses the right of wilderness and wild species to exist regardless of their present or potential future value to man. Rolston states that to value all other species only for human interests is like a nation's arguing all its foreign policy in terms of national interest. Neither seems fully moral. He argues that there is something overspecialized about an ethic held by the dominant class of *Homo sapiens* that regards the welfare of only one of several million species as an object of duty. There is something morally naive about living in a reference frame wherein one species takes itself as absolute and values everything else only in terms of utility to that one species. He challenges man to accept interspecific as well as intraspecific altruism.

The very biblical verse quoted above and used to justify man's right to exploit other species can be interpreted as a mandate for man to practice interspecific altruism. The verse gives man dominion over other species, but the right of dominion, that is the right of ownership or authority, does not necessarily include the right to destroy. In fact, the giving of "dominion" could be interpreted as a mandate of responsibility to practice good stewardship, that is, to care for, and to preserve from destruction, all the species of the earth. Species destruction would be a violation of this mandate.

Future Directions

In the introduction to this book, the pattern of development and opening up of the Amazon frontier was compared to the opening of the North American west over 100 years ago. In both cases, it appears that a valuable natural resource, the native forest, was not used as wisely and efficiently as might have been possible. It is difficult to criticize countries with territories in the Amazon for hurried and wasteful development. After all, don't their citizens and governments have the right to exploit and develop the wilderness in whatever manner they wish? The North Americans certainly claimed that right for their frontier. Even today, few would deny a nation the right to develop its resources.

However, the right to exploit and develop the Amazon is not the question for this book. Rather, the question is, given the pressure for development in the Amazon, what are the best land uses and management alternatives? Just because the original North American land and forest resources appear not to have been used well from today's modern perspective does not mean that the mistakes must be repeated. To repeat in South America the wasteful destruction of forests that occurred in North America would be to say that Civilized Western Man has not learned anything or improved his sense of awareness in over 100 years. Certainly an effort should be made to use the Amazon forests more wisely.

Comparison of the case studies, and their evaluation in light of other development systems, suggests that the best system for the Amazon region in terms of ecological and economic sustainability is a system with a high-intensity concentration of cultural production factors per unit area of land. Because high-intensity concentration is often on a small scale it is environmentally desirable, because concentration in small areas uses less land and allows preservation of larger areas of undisturbed forest. It is energetically desirable, because it has a high energy use efficiency compared to large-scale cutting of the forest, which does not efficiently use the energetic services of the forest. Economically it is desirable because a relatively high concentration of production factors on small units of land results in a relatively high geographic concentration of productivity and profits. Concentration of agricultural infrastructure leads to a greater sustainability of agricultural output, and this in turn leads to greater social stability. Socially stable regions are more readily incorporated into the national political framework of a country.

This does not mean that large-scale projects cannot be economically profitable, or socially desirable. It means, though, that large-scale of development must be accompanied by high concentration of intensity per unit land. High total investment of capital and labor resources in a project will not alone result in a successful project. To achieve the efficiency and economy of scale, the scale must be large for each unit of land.

Examples of systems with intensely concentrated factors of production include the recently initiated intensive cultivation with mulches described for Yurimaguas by Wade and Sanchez (1983). In other examples, labor is applied in intense concentration as in agroforestry. Agroforestry combines elements of managed fallow and native species utilization, described in the Gran Pajonal and sustained-yield forestry cases, with labor-intensive perennial cropping systems, such as at Tome-Assú. Agroforestry systems can include production of a variety of annual and perennial crops, with trees for wood, fuel, fruit, and extractants, such as rubber and oils.

In agroforestry systems, nature's services are used efficiently, and consequently social and economic welfare is greater at a lower cost. Agroforestry takes better advantage of the naturally occurring services of nature than other sytems more heavily dependent on cultural energy supplements. Agroforestry systems more closely resemble the naturally occurring ecosystems of the Amazon than other systems, such as pasture or grain monocultures, and consequently are better adapted to preventing nutrient loss and soil erosion and compaction. When the energetic services of nature and the energy value of labor are added to any fossil fuel subsidies that agroforestry systems may receive, they may have a relatively high-intensity concentration of energy per unit land, compared to most other systems.

In conclusion, the most desirable systems for development of the Amazon, from a variety of viewpoints, are those best able to take advantage of the naturally occurring resources of the rain forest. In the long term such systems function better, are more profitable, more sustainable, and more socially and politically beneficial, and are less environmentally damaging than systems that have evolved in other regions and under other cultural and environmental conditions.

References

Aubertin, G. M., and J. H. Patric. 1974. Water quality after clearcutting a small watershed in West Virginia. *Journal of Environmental Quality* 3:243–249.

Baur, G. N. 1964. *The Ecological Basis of Rainforest Management*. New South Wales, Australia: Forestry Commission.

Bayliss-Smith, T. P. 1982. The ecology of agricultural systems. Cambridge: Cambridge University Press.

Black, C. A., ed. 1965. *Methods of Soil Analysis*. Madison, Wisconsin: American Society of Agronomy.

Bormann, F. H., G. E. Likens, D. W. Fisher, and R. S. Pierce. 1968. Nutrient loss accelerated by clear-cutting of a forest ecosystem. *Science* 159:882–884.

Braun-Blanquet, J. 1965. *Plant Sociology*. New York: Hafner.

Brinkmann, W. L. F., and J. C. Nascimento. 1973. The effect of slash and burn agriculture on plant nutrients in the tertiary region of Central Amazonia. *Acta Amazonica* 3:55–61.

Buschbacher, R. J. 1984. *Changes in Productivity and Nutrient Cycling following Conversion of Amazon Rainforest to Pasture*. Ph.D. Dissertation. Athens, Georgia: Institute of Ecology, University of Georgia.

——, C. Uhl, and E.A.S. Serrão. 1984. Forest development following pasture use in the North of Pará, Brazil. In *First Symposium on Development in the Humid Tropics*. M. Dantas ed. Belém, Brazil: Empresa Brasileira de Pesquisa Agropecuaria.

Cain, S. and G. M. de Oliveira Castro. 1971. *Manual of Vegetation Analysis*. New York: Hafner.

Campbell, J. R., and J. F. Lasley. 1975. *The Science of Animals That Serve Mankind*. New York: McGraw-Hill Publications in the Agricultural Sciences.
Chijicke, E. O. 1980. *Impact on Soils of Fast-Growing Species in Lowland and Humid Tropics*. FAO Forestry Paper No. 21. Rome: Food and Agriculture Organization of the United Nations.
Christianson, R. A. 1984. Energy Perspectives on a Tropical Forest/Plantation System at Jari, Brazil. M. S. thesis, University of Florida, Gainesville, Florida.
Clark, C., and M. Haswell. 1970. *The Economics of Subsistence Agriculture*. London: Macmillan.
Cochrane, T. T. and P. A. Sanchez. 1982. Land resources, soils and their management in the Amazon region: a state of knowledge report. *Amazonia: Agriculture and Land Use Research*. Cali, Colombia: Centro Internacional de Agricultura Tropical, pp. 135–210.
Cole, D. W., and C. M. Bigger. Undated. *Effect of harvesting and residue removal on nutrient losses and productivity*. Fifth Annual Report. College of Forest Resources, University of Washington, Seattle, Washington.
Conklin, H. C. 1959. Population-land balance under systems of tropical forest agriculture. *Proceedings of the ninth Pacific Science Congress* 7:63.
Conway, G. R. 1983. *Agroecosystem Analysis*. Imperial College Centre for Environmental Technology Series E, No. 1. 48 Prince's Gardens, London: Imperial College of Science and Technology.
Costanza, R. 1980. Embodied energy and economic valuation. *Science* 210:1219–1224.
Cottam, G., and J. T. Curtis. 1956. The use of distance measures in phytosociological sampling. Ecology 37:451–60.
Cox, G. W., and M. D. Atkins. 1979. *Agricultural Ecology*. San Francisco: Freeman.
Davis, S. H. 1977. *Victims of the Miracle*. Cambridge: Cambridge University Press.
de Graaf, N. R. 1982. Sustained timber production in the tropical rainforest of Suriname. In *Management of Low Fertility Acid Soils of the American Humid Tropics*. J. F. Wienk and H. A. de Wit, eds. Instituto Internacional de Ciencias Agricolas, San Jose, Costa Rica: pp. 175–189.
de La Marca, J. B. 1973. Diaria y padron de la gente del Pajonal. Ms., Biblioteca Nacional, Lima, Peru.
De Melo Carvalho, J. C. 1984. The conservation of nature in the Brazilian Amazonia. *In* The Amazon; Limnology and Landscape Ecology of a Mighty Tropical River and Its Basin. H. Sioli, ed. Dordrecht: Junk, pp. 707–736.
Denevan, W. M. 1971. Campa subsistence in the Gran Pajonal, Eastern Peru. *Geographical Review* 61:496–518.
——, J. M. Treacy, J. B. Alcorn, C. Padoch, J. Denslow, and S. Flores Paitan. 1984. Indigenous agroforestry in the Peruvian Amazon: Bora Indian managment of swidden fallows. *Interciencia* 9:346–357.
de Oliveira, A. E. 1983. Ocupacao humana. In *Amazonia: Desenvolvimento, Integracao e Ecologia*. Conselho Nacional de Desenvolvimento Cientifico e Technologico. Sao Paulo, Brazil: Editora Brasiliense S. A. pp. 144–127.
de Schlippe, P. 1956. *Shifting cultivation in Africa*. London: Routledge and Kegan Paul.
Dubroeucq, D., and V. Sanchez. 1981. *Caracteristicas Ambientales y Edaficas del Area Muestra San Carlos de Rio Negro-Solano*. Caracas, Venezuela: Ministerio del Ambiente y de los Recursos Naturales Renovables, Serie Informe Cientifico DGSIIA/IC/12.
Falesi, I. 1976. Ecosistema de pastagem cultivada na Amazonia Brasileira. Boletim Tecnico I CPATU. Belém, Para: EMBRAPA.
Fearnside, P. M. 1978. *Estimation of Carrying Capacity for Human Populations in a Part of the Transamazon Highway Colonization Area of Brasil*. Ph.D. dissertation. Ann Arbor, Michigan: Dept. of Biological Sciences, University of Michigan.
—— 1979. Cattle yield prediction for the Transamazon highway of Brazil. *Interciencia* 4(4):220–226.

—— and J. M. Rankin. 1980. Jari and development in the Brazilian Amazon. *Interciencia* 5(3):146–156.
—— and —— 1982. The new Jari: Risks and prospects of a major Amazon development. *Interciencia* 7(6):329–339.
—— and —— 1985. Jari revisited: changes and the outlook for sustainability in Amazonia's largest silvicultural estate. *Interciencia* 10(3):121–129.
Feller, M. C., and J. P. Kimmins. 1984. Effects of clearcutting and slash burning on streamwater chemistry and watershed nutrient budgets in southwestern British Columbia. *Water Resources Research* 20:29–40.
Fluck, R. C., and C. D. Baird. 1980. *Agricultural energetics.* Westport, Conn: AVI Publishing Co.
Foweraker, J. 1981. *The Struggle for Land: A Political Economy of the Pioneer Frontier in Brazil from 1930 to the Present Day.* Cambridge: Cambridge University Press.
Fox, J. E. D. 1976. Constraints on the natural regeneration of tropical moist forest. *Forest Ecology and Management* 1:37–65.
Freeman, J. D. 1955. Iban agriculture: a report on the shifting cultivation of hill rice by the Iban of Sarawak. *Colonial Research Studies, No. 18.* London: H. M. S. O.
Fritzell, P. A. 1983. Changing conceptions of the Greak Lakes forest: Jacques Cartier to Sigurd Olson. In *The Great Lakes Forest. An environmental and social history.* S. L. Flader, ed. Minneapolis, Minn.: Univ. of Minnesota Press, pp. 274–294.
Gall, N. 1979. Ludwig's Amazon empire. *Forbes* May 14, 1979:127–144.
Gebhardt, M. R., T. C. Daniel, E. E. Schweizer, and R. R. Allmaras. 1985. Conservation tillage. *Science* 230:625–630.
Gessel, S. P., and D. W. Cole. 1965. Influence of removal of forest cover on movement of water and associated elements through soil. *Journal of the American Water Works Association.* 57:1301–1310.
Gilliland, M. W. 1975. Energy analysis and public policy. *Science* 189:1051–1056.
Goodland, R. J. A. 1980. Environmental ranking of Amazonian development projects in Brazil. *Environmental Conservation* 7:9–26.
Goodland, R. J. A., and H. S. Irwin. 1975. *Amazon Jungle: Green Hell to Red Desert?* Amsterdam: Elsevier.
Greaves, A. 1979. Gmelina large scale planting, Jarilandia, Amazon basin. *Commonwealth Forestry Review* 58:267–269.
Gross, D. R., G. Eiten, N. M. Flowers, F. M. Leoi, M. L. Ritter, and D. W. Werner. 1979. Ecology and acculturation among native peoples of central Brazil. *Science* 206:1043–1050.
Hall, C. W. 1984. The role of energy in world agriculture and food availability. In *Food and Energy Resources.* D. Pimentel and C. W. Hall eds. Orlando, Florida: Academic Press, pp. 43–63.
Harris, D. R. 1971. The ecology of swidden cultivation in the upper Orinoco rain forest, Venezuela. *The Geographical Review* 61:475–495.
Hartshorn, G. 1979. *Report on Activities as a Forest and Man Fellow to Institute of Current World Affairs, Sept. 1979.* Sponsored by Institute of Current World Affairs, Wheelock House, Hanover, New Hampshire.
—— 1981. *Report to Institute of Current World Affairs, December, 1981, on Activities as a Forest and Man Fellow, Sponsored by That Institute.* Institute of Current World Affairs, Wheelock House, Hanover, New Hampshire.
Hecht, S. B. 1984. Cattle ranching in Amazonia: political and ecological considerations. In *Frontier Expansion in Amazonia.* M. Schmink and C. H. Wood, eds. Gainesville, Florida: University of Florida Press, pp. 366–398.
Heichel, G. H. and C. R. Frink. 1975. Anticipating the energy needs of American agriculture. *Journal of Soil and Water Conservation* 30:48–53.
Herrera, R., C. F. Jordan, E. Medina, and H. Klinge. 1981. How human activities disturb the nutrient cycles of a tropical rainforest in Amazonia. *Ambio* 10:109–114.
Holden, C. 1986. Environment and development. *Science* 231:1494.

Hornbeck, J. W., and W. Kropelin. 1982. Nutrient removal and leaching from a whole-tree harvest of northern hardwoods. *Journal of Environmental Quality* 11:309–316.
Hornick, J. R., J. I. Zerbe, and J. L. Whitmore. 1984. Jari's successes. *Journal of Forestry* 82(11):663–670.
Huettner, D. A. 1976. Net energy analysis: an economic assessment. *Science* 192:101–104.
Hyman, E. L. 1980. Net energy analysis and the theory of value: is it a new paradigm for a planned economic system? *Journal of Environmental Systems* 9:313–324.
Jonkers, W. B. J. and P. Schmidt. 1984. Ecology and timber production in tropical rainforest in Suriname. *Interciencia* 9:290–297.
Jordan, C. F. 1971. A world pattern in plant energetics. *American Scientist* 59:425–433.
—— 1980. Nutrient dynamics of a tropical rain forest ecosystem, and changes in the nutrient cycle due to cutting and burning. Annual Report to the National Science Foundation. Athens, Georgia: Institute of Ecology.
—— 1982. Amazon rain forests. *American Scientist* 70:394–401.
—— 1985a. *Nutrient Cycling in Tropical Forest Ecosystems*. Chichester: Wiley.
—— 1985b. Jari: A development project for pulp in the Brazilian Amazon. *The Environmental Professional* 7:135–142.
——, and C. Uhl. 1978. Biomass of a "tierra firme" forest of the Amazon Basin. *Oecologia Plantarum* 13:387–400.
——, and R. Herrera. 1981. Tropical rain forests: are nutrients really critical? *American Naturalist* 117:167–180.
——, F. Golley, J. B. Hall and J. Hall. 1980. Nutrient scavenging of rainfall by the canopy of an Amazonian rain forest. *Biotropica* 12:61–66.
Key, M. R. 1979. *The grouping of South American Indian Languages*. Tubingen: Gunter Narr Verlag.
Kimmins, J. P. 1977. Evaluation of the consequences for future tree productivity of the loss of nutrients in whole-tree harvesting. *Forest Ecology and Management* 1:169–183.
Kinkead, G. 1981. Trouble in D. K. Ludwig's jungle. *Fortune* April 20, 1981:102–117.
Korten, D. C. and R. Klauss. 1984. *People-Centered Development*. West Hartford, Connecticut: Kumarian Press.
Kyrklund, B. 1976. Paper from mixed tropical forests. *Unasylva* 28(112–113):86–92.
Lathrup, D. 1970. *The upper Amazon*. New York: Praeger Press.
Leach, G. 1975. Net energy analysis—is it any use? *Energy Policy* 3:332–344.
Lowe, R. G. 1977. Experience with the tropical shelterwood system of regeneration in natural forest in Nigeria. *Forest Ecology and Management* 1:193–212.
Lowrance, R., B. R. Stinner, and G. J. House. 1984. *Agricultural Ecosystems*. New York: Wiley.
Lugo, A. E., and S. Brown. 1982. Conversion of tropical moist forests: a critique. *Interciencia* 7:89–93.
Masello, R. 1979. The last billionare. *Clipper* (Pan American Airlines), September 1979: 54–57.
McNaughton, S. J. 1976. Serengeti migratory wildebeest: facilitation of energy flow by grazing. *Science* 191:92–94.
Meggers, B. J. 1971. *Amazonia, Man and Culture in a Counterfeit Paradise*. Chicago: Aldine-Atherton.
—— 1985. Aboriginal adaptation to Amazonia. In *Amazonia*. G. T. Prance and T. E. Lovejoy eds. Key Environments Series. Oxford: Pergamon Press, pp. 307–327.
Miller, D. 1983. Entrepreneurs and bureaucrats: the rise of an urban middle class. In *The dilemma of Amazonian development*. E. F. Moran ed. Boulder, Colorado: Westview, pp. 65–93.
Miller, J. H., and M. Newton. 1983. Nutrient loss from disturbed forest watersheds in Oregon's coast range. *Agro-Ecosystems* 8:158–167.
Minson, D. J. 1980. Nutritional differences between tropical and temperate pastures. In

Grazing Animals. F. H. W. Morley ed. Amsterdam: Elsevier Publishing Co., pp. 143–157.

Miracle, M. P. 1973. The Congo Basin as a habitat for man. In *Tropical Forest Ecosystems in Africa and South America: A Comparative Review*. B. J. Meggers, E. S. Ayensu, and W. D. Duckworth eds. Washington, D.C.: Smithsonian Institution Press, pp. 335–344.

Molion, L. C. B. 1975. *A Climatonomic Study of the Energy and Moisture Fluxes of the Amazonian Basin with Considerations of Deforestation Effects*. Ph.D. dissertation, University of Wisconsin, Madison, Wisconsin.

Monk, C. D. 1975. Nutrient losses in particulate form as weir pond sediments from four unit watersheds in the southern Appalachians. In *Mineral Cycling in Southeastern Ecosystems*. F. G. Howell, J. B. Gentry, and M. H. Smith eds. Washington, D.C.: Energy Research and Development Administration CONF 740-513. pp. 862–867.

Montagnini, F., and C. F. Jordan. 1983. The role of insects in the productivity decline of cassava (*Manihot esculenta* Crantz) on a slash and burn site in the Amazon Territory of Venezuela. *Agriculture, Ecosystems, and Environment* 9:293–301.

Moran, E. F. 1981. *Developing the Amazon*. Bloomington, Indiana: Indiana University Press.

—— 1984. Colonization in the Transamazon and Rondonia. In *Frontier Expansion in Amazonia*. M. Schmink and C. H. Wood eds. Gainesville, Florida: University of Florida Press. pp. 285–303.

—— and J. Hill. 1982. The cultural ecology of the Rio Negro Basin. Paper presented at the Symposium on The Structure and Function of Amazonian Forest Ecosystems of the Upper Rio Negro, Caracas, Venezuela, Nov. 8–12, 1982.

Mueller-Dombois, D., and H. Ellenberg. 1974. *Aims and Methods of Vegetation Ecology*. New York: Wiley.

Myers, N. 1980. *Conversion of Tropical Moist Forests*. Washington, D.C.: National Academy of Sciences. Pub. ISBN 0-309-02945-7

—— 1983. *A Wealth of Wild Species: Storehouse for Human Welfare*. Boulder, Colorado: Westview Press.

—— 1984. *The Primary Source: Tropical Forests and Our Future*. New York: Norton.

Nascimento, C., and A. Homma. 1984. *Amazonia: Meio Ambiente e Tecnologia Agricola*. EMBRAPA-CPATU Documentos *27*. Caixa Postal 48, Belém, Pará, Brazil.

Norman, M. J. T. 1978. Energy inputs and outputs of subsistence cropping systems in the tropics. *Agro-Ecosystems* 4:355–366.

North Carolina State University 1972, 1973, 1974, 1976–1977, 1978–1979, 1980–1981. *Tropical Soils Research Program Annual Reports*. Raleigh, North Carolina: Soil Science Dept., North Carolina State University.

Nye, P. H. and D. J. Greenland. 1960. *The Soil under Shifting Cultivation*. Tech. Comm. 51, Commonwealth Agricultural Bureaux, Farnham Royal, Bucks. England.

—— and —— 1964. Changes in the soil after clearing tropical forest. *Plant and Soil* 21:101–112.

Odend'hal, S., R. C. De, and S. K. Dutta. 1979. A note on the estimation of weight of adult Bengali cows by the Minnesota formula. *Indian Journal of Animal Science* 49:852–853.

Odum, H. T. 1983. *Systems Ecology*. New York: Wiley Interscience.

—— and E. C. Odum. 1981. *Energy Basis for Man and Nature*. New York: McGraw-Hill Book Co.

Ogawa, H., K. Yoda, and T. Kira. 1961. A preliminary survey on the vegetation of Thailand. *Nature and Life in Southeast Asia* 1:21–157.

Okigbo, B. N. 1984. Improved permanent production systems as an alternative to shifting intermittent cultivation. In *Improved Production Systems as an Alternative to Shifting Cultivation*. FAO Soils Bulletin #53. Rome: FAO. pp. 1–100

Pearsall, S. H. 1984. *In Absentia* benefits of nature preserves: a review. *Environmental Conservation* 11:3–10.

Pharmacy Times Journal, Sept. 1980. pp. 106–109.
Pierce, R. S., C. W. Martin, C. C. Reeves, G. E. Likens, and F. H. Bormann. 1972. Nutrient loss from clearcuttings in New Hampshire. In *Symposium on Watersheds in Transition.* Ft. Collins, Colorado: Colorado State University. pp 285–295.
Pimentel, D. 1980. *Handbook of Energy Utilization in Agriculture.* Boca Raton, Florida: CRC Press.
——, and M. Burgess. 1980. Energy inputs in corn production. In *Handbook of Energy Utilization in Agriculture.* D. Pimentel, ed. Boca Raton, Florida: CRC Press. pp. 67–84.
——, L. E. Hurd, A. C. Bellotti, M. J. Forster, I. N. Oka, O. D. Sholes, and R. J. Whitman. 1973. Food production and the energy crisis. *Science* 182:443–449.
Poore, D. 1976. The values of the tropical moist forest ecosystems and the environmental consequences of their removal. Item 8(a) of the Provisional Agenda, Committee on Forest Development in the Tropics, Fourth Session, Rome, Italy, 15–20 Dec. 1976. FAO Pub. FO:FDT/76/8(a). Rome: FAO.
Posey, C. E. 1980. Statement of Dr. Clayton E. Posey on Jari Florestal. In Hearings before the Subcommittee on Foreign Affairs, House of Representatives, 96th Congress, Second Session, May 7, June 19, and September 18, 1980: Tropical Deforestation. Washington, D.C.: U.S. Government Printing Office, pp. 428–433.
Prance, G. T. 1982. *Biological Diversification in the Tropics.* New York: Columbia Univ. Press.
Rankin, J. R. 1985. Forestry in the Brazilian Amazon. In *Amazonia.* G. T. Prance, and T. E. Lovejoy, eds. Key Environments series. Oxford: Pergamon Press. pp. 369–392.
Rappaport, R. A. 1971. The flow of energy in an agricultural society. *Scientific American* 224 (3):117:116–132.
Reinhardt, H. H. 1983. Social adjustments to a changing environment. In *The Great Lakes Forest. An Environmental and Social History.* S. L. Flader, ed. Minneapolis, Minnesota: University of Minnesota Press, pp. 205–219.
Rich, B. M. 1985. The multilateral development banks, environmental policy, and the United States. *Ecology Law Quarterly* 12(4):681–745.
Rolston, H. 1985. Duties to endangered species. *BioScience* 35:718–726.
Russell, C. E. 1983. *Nutrient Cycling and Productivity of Native and Plantation Forests at Jari Florestal, Pará, Brazil.* Ph.D. dissertation. Athens, Georgia: Institute of Ecology, University of Georgia.
Rutger, J. N., and W. R. Grant. 1980. Energy use in rice production. In *Handbook of Energy Utilization in Agriculture.* D. Pimentel, ed. Boca Raton, Florida: CRC Press, pp. 93–98.
Salati, E. and P. B. Vose. 1984. Amazon Basin: a system in equilibrium. Science 225:129–138.
Sanchez, P. A. 1976. *Properties and Management of Soils in the Tropics.* New York: Wiley.
—— 1981. Soils of the humid tropics. In *Blowing in the Wind: Deforestation and Long-Range Implications.* V. H. Sutlive, N. Altshuler, and M. D. Zamora eds. Studies in Third World Societies No. 14, Department of Anthropology, College of William and Mary, Williamsburg, Virginia, pp. 347–410.
——, D. E. Bandy, J. H. Villachica, and J. J. Nicholaides. 1982. Amazon basin soils: management for continuous crop production. *Science* 216:821–827.
——, J. H. Villachica, and D. E. Bandy. 1983a. Soil fertility dynamics after clearing a tropical rainforest in Peru. *Soil Science Society of America Journal* 47:1171–1178.
——, D. E. Bandy, J. H. Villachica, and J. J. Nicholaides. 1983b. Continuous cultivation and nutrient dynamics. pp 11–16 *in* J. J. Nicholaides, W. Couto and M. K. Wade eds. *1980–1981 Technical Report, Agronomic-Economic Research on Soils of the Tropics.* Soil Science Dept., North Carolina State University, Raleigh, N.C.
Savage, J. M., C. R. Goldman, D. P. Janos, A. E. Lugo, P. H. Raven, P. A. Sanchez,

and H. G. Wilkes. 1982. *Ecological Aspects of Development in the Humid Tropics*. Washington, D.C.: National Academy Press.

Sawyer, D. R. 1984. Frontier expansion and retraction in Brazil. In *Frontier Expansion in Amazonia*. M. Schmink and C. H. Wood eds. Gainesville, Florida: University of Florida Press. pp. 180–203.

Serrão, E. A. S., I. C. Falesi, J. Bastos de Veiga, and J. F. Teixeira Neto. 1979. Productivity of cultivated pastures on low fertility soils in the Amazon of Brazil. In *Pasture Production in Acid Soils of the Tropics*. P. A. Sanchez and L. E. Tergas, eds. Cali, Colombia: Centro Internacional de Agricultura Tropical, pp. 195–225.

Shubart, H. O. R. 1983. Ecologia e utilizacao das florestas. In *Amazonia: Desenvolvimento, Integracao e Ecologia*. Conselho Nacional de Desenvolvimento Cientifico e Technologico. Sao Paulo, Brazil: Editora Brasiliense S. A. pp. 101–143.

Sioli, H. 1973. Recent human activities in the Brazilian Amazon region and their ecological effects. In *Tropical Forest Ecosystems in Africa and South America: A Comparative Review*. B. J. Meggers, E. S. Ayensu, and W. D. Duckworth eds. Washington, D.C.: The Smithsonian Institution. pp. 321–344.

Smith, N. J. H. 1978. Agricultural productivity along Brazil's transamazon highway. *Agro-Ecosystems* 4:415–432.

—— 1981. Colonization lessons from a tropical forest. *Science* 214:755–761.

—— 1982. *Rainforest Corridors, the Transamazon Colonization Scheme*. Berkeley: University of California Press.

Sollins, P., and F. M. McCorison. 1981. Nitrogen and carbon solution chemistry of an old growth coniferous forest watershed before and after cutting. *Water Resources Research* 17:1409–1418.

Stark, N., and M. Spratt. 1977. Root biomass and nutrient storage in rain forest Oxisols near San Carlos de Rio Negro. *Tropical Ecology* 18:1–9.

Steinhart, C. E. and J. S. Steinhart. 1974. *Energy: Sources, Use and Role in Human Affairs*. Belmont, California: Duxbury Press, Wadsworth Pub. Co.

Stewart, G. A. 1970. High potential productivity of the tropics for cereal grasses, grass forage crops, and beef. *Journal of the Australian Institute of Agricultural Science* 36:85–101.

Stone, R. D. 1985. *Dreams of Amazonia*. New York: Viking.

Swank, W. T., and J. E. Douglass. 1975. Nutrient flux in undisturbed and manipulated forest ecosystems in the southern Appalachian mountains. In Publication No 117 de l' Association Internationale des Sciences Hydrologiques Symposium de Tokyo. pp. 445–456.

——, and ——. 1977. Nutrient budgets for undisturbed and manipulated hardwood forest ecosystems in the mountains of North Carolina. In *Watershed Research in Eastern North America*. D. L. Correll, ed. Edgewater, Maryland: Chesapeake Bay Center for Environmental Studies. pp 343–364.

——, and J. B. Waide. 1980. Interpretation of nutrient cycling research in a management context: evaluating potential effects of alternative management strategies on site productivity. In *Forests: Fresh Perspectives from Ecosystem Analysis*. Proceedings 40th Annual Biology Colloquium. Corvallis, Oregon: Oregon State University Press, pp. 137–158.

Synnot, T. J. and R. H. Kemp. 1976. Choosing the best silvicultural system. *Unasylva* 28(112–113):74–79.

Tamm, C. O. 1979. Nutrient cycling and productivity of forest ecosystems. In *Impact of Intensive Harvesting on Forest Nutrient Cycling*. A. L. Leaf, program chairman. Proceedings of a Symposium at Syracuse, New York, August 13–15, 1979. Pub. Northeast Forest Experiment Station, Broomall, Pennsylvania. pp. 2–21.

Teeter, R. E., F. N. Owens, and G. W. Horn. 1979. Ytterbium as a ruminal marker. *Abstracts American Society of Animal Science*, p. 412.

Teitzel, J. K., A. R. McTaggart, and M. J. Hibberd. 1971. Pasture and cattle management in the wet tropics. *Queensland Agricultural Journal* 97:25–29.

Time, 1976. Ludwig's wild Amazon kingdom. November 15, 1976:59–59A.
——. 1979. Billionaire Ludwig's Brazilian gamble. September 10, 1979:76–78.
——, 1982. End of a billion-dollar dream. January 25, 1982:59.
Toledo, J. M., and V. A. Morales. 1979. Establishment and management of improved pasture in the Peruvian Amazon. In *Pasture Production in Acid Soils of the Tropics.* P. A. Sanchez and L. E. Tergas eds. Cali, Colombia: Centro Internacional de Agricultura Tropical, pp. 177–194.
Twining, C. E. 1983. The lumbering frontier. In *The Great Lakes Forest. An Environmental and Social History.* S. L. Flader, ed. Minneapolis, Minnesota: University of Minnesota Press, pp. 121–136.
Uhl, C. 1982. Rio Negro forest perturbations: the recovery process. Paper presented at the 1982 symposium, Structure and function of Amazonian Forest Ecosystems in the Upper Rio Negro. Caracas, Venezuela, November 1982.
——, and R. Buschbacher. 1985. A disturbing synergism between cattle ranch burning practices and selective tree harvesting in the Eastern Amazon. *Biotropica* 17:265–268.
——, and C. F. Jordan. 1984. Vegetation and nutrient dynamics during the first five years of succession following forest cutting and burning in the Rio Negro region of Amazonia. *Ecology* 65:1476–1490.
——, and P. Murphy. 1981. A comparison of productivities and energy values between slash and burn agriculture and secondary succession in the upper Rio Negro region of the Amazon Basin. *Agro-Ecosystems* 7:63–83.
UNESCO, 1983. *Swidden Cultivation in Asia. I. Content Analysis of the Existing Literature: A Stocktaking Exercise.* Bangkok, Thailand: UNESCO Regional Office for Education in Asia and the Pacific.
van Beukering, J. A. 1947. Het Ladangvraagstak. *Landbouw Buitenzorg* 29:241–285.
Vicente-Chandler, J. 1974. Fertilization of humid tropical grasslands. In *Forage Fertilization.* D. A. Mays. ed. Madison, Wisconsin: American Society of Agronomy. pp. 277–300.
——, R. Caro-Costas, R. W. Pearson, F. Abruna, J. Figarella, and S. Silva. 1964. *The Intensive Management of Tropical Forages in Puerto Rico.* Rio Piedras, Puerto Rico: University of Puerto Rico Agricultural Experiment Station Bulletin 187.
Villachica, J. H., C. E. Lopez, and P. A. Sanchez. 1976. Continuous cropping experiment. In *Agronomic-Economic Research on Tropical Soils, Annual Report for 1975.* Raleigh, North Carolina: Soil Science Department, North Carolina State University, pp. 117–137.
Wade, M. K., and P. A. Sanchez, 1983. Mulching and green manure applications for continuous crop production in the Amazon Basin. *Agronomy Journal* 75:39–45.
Wadsworth, F. H. 1983. Production of usable wood from tropical forests. In *Tropical Rain Forest Ecosystems. Ecosystems of the World,* Vol. 14A. F. B. Golley, Vol. Ed. Amsterdam: Elsevier, pp 279–288.
Wagley, C. 1984. Foreword. In *Frontier Expansion in Amazonia.* M. Schmink and C. H. Wood eds. Gainesville, Florida: University of Florida Press, pp. ix–xiv.
Watters, R. F. 1971. *Shifting Cultivation in Latin America.* FAO forestry development paper No. 17. Rome: FAO.
Webb, M., and D. Pearce. 1975. The economics of energy analysis. *Energy Policy* 3(4):318–331.
Webster, J. R., and B. C. Patten. 1979. Effects of watershed perturbation on stream potassium and calcium dynamics. *Ecological Monographs* 49:51–72.
Westman, W. E. 1977. How much are nature's services worth? *Science* 197:960–964.
Whicker, F. W., and V. Schultz. 1982. *Radioecology: Nuclear Energy and the Environment I.* Boca Raton, Florida: CRC Press.
White, L. 1967. The historical roots of our ecologic crisis. *Science* 115:1203–1207.
Wilbert, J. 1972. *Survivors of Eldorado.* New York: Praeger.
Woessner, R. A. 1982. Plantation forestry and natural forest utilization in the Amazon

Basin. Paper presented at the American Society of Foresters meeting, September 19–22, Cincinnati, Ohio.

Wood, C. H., and J. Wilson. 1984. The magnitude of migration to the Brazilian frontier. In *Frontier Expansion in Amazonia*. M. Schmink and C. H. Wood eds. Gainesville, Florida: University of Florida Press, pp. 142–152.

Zinke, P. J., S. Sabhasri, and P. Kunstadter. 1978. Soil fertility aspects of the Lua' forest fallow system of shifting cultivation. In *Farmers in the Forest*. P. Kunstadter, E. C. Chapman, and S. S. Sabhasri, eds. Honolulu: University of Hawaii. pp 134–159.

Index